New Advances in Polyolefins

New Advances in Polyolefins

Edited by

T. C. Chung
The Pennsylvania State University
University Park, Pennsylvania

Springer Science+Business Media, LLC

Library of Congress Cataloging-in-Publication Data

New advances in polyolefins / edited by T. C. Chung.
        p.   cm.
      "Proceedings of an ACS symposium on new advances in polyolefins
   polymers, held August 23-28, 1992, in Washigton, D.C."--T.p. verso.
   Includes bibliographical references and index.
      ISBN 978-0-306-44588-0   ISBN 978-1-4615-2992-7 (eBook)
      DOI 10.1007/978-1-4615-2992-7
      1. Polyolefins--Congresses.   I. Chung, T. C.
   TP1180.P68N49 1994
   668.4'234--dc20   93-38227
                                                               CIP

Proceedings of an ACS symposium on New Advances in Polyolefins Polymers, held August 23–28, 1992, in Washington, D.C.

ISBN 978-0-306-44588-0

© 1993 Springer Science+Business Media New York
Originally published by Plenum Press New York in 1993

PREFACE

Polyolefin is one of the most important materials produced in the chemical industry. The research in this area is not only scientifically challenging but also potentially economically rewarding. Many research activities, such as developing new catalysts, understanding polymerization mechanisms, modifying the products and improving the physical properties of the material, have been proceeding at a very fast pace, especially in the industrial laboratories in many countries.

It is very important and exciting to bring researchers active in this area, from both the academic and industrial sectors, to communicate their new findings. To the best of my knowledge, a symposium covering diverse aspects of polyolefin research has not been held in the USA for many years. With this in mind, a symposium entitled "Recent Advances in Polyolefins" was held at the National ACS Meeting in Washington D.C., August 23-28, 1992. This symposium covered both scientific and technological aspects of polyolefin, which included four sections: Catalyst/Polymerization, Functionalization of Polyolefins, Blends of Polyolefins with Other Polymers, and Applications. More than 50 papers, including 20 foreign contributions, were presented during three and a half days of meetings. It was truly exciting to see key researchers from both academic and industrial laboratories exchange their recent results and to share the important developments in polyolefins.

This volume is based on the same spirit and is intended to capture some of the most recent and emerging technical achievements presented at the meeting. It is certainly impossible to cover every aspect in this rapidly expanding research area. Instead, the chapters in this volume provide exciting results and balanced coverage of research in various areas: Catalyst/Polymerization, Functionalization, Characterization and Polyolefin Blends, and Composites. This book is useful as a reference for scientists interested in the recent developments of polyolefins.

It was both a pleasure and an honor to be the symposium chairman and editor of these proceedings. I thank the authors for their fine contributions. I also extend my thanks to the following organizations for their generous financial support:

ACS, Division of Polymeric Materials Science and Engineering
Petroleum Research Foundation
Ethyl Corporation
Exxon Research and Engineering Co.
Mobil Chemical Company
Texas Eastman Company

Most of the funds were used to enable some of the foreign speakers to participate in this meeting.

T.C. (Mike) Chung
Pennsylvania State University
University Park, PA 16802

CONTENTS

CATALYST/POLYMERIZATION

FUNCTIONALIZATION

CHARACTERIZATION

POLYOLEFIN BLENDS AND COMPOSITES

MODEL SILICA SUPPORTED OLEFIN POLYMERIZATION CATALYSTS

Jonathan P. Blitz

Quantum Chemical Corporation
USI Division, Basic Research
11530 Northlake Drive
Cincinnati,OH 45249

ABSTRACT

Chemically modified silica surfaces have been synthesized and characterized as model olefin polymerization catalysts. By variation of silica gel pretreatments, $TiCl_4$ can be reacted with various surface bonding geometries. Most surface characterization was done using diffuse reflectance FTIR spectroscopy. The $TiCl_4$ reacted silica gels, after reaction with $Al(Et)_3$, give catalysts which polymerize ethylene and 1-butene with different activities and comonomer incorporation.

INTRODUCTION

Ziegler-Natta catalysts based on titanium chlorides and aluminum alkyls are extremely important for the polymerization of alpha-olefins. Originally these heterogeneous catalysts were made by the reaction of $TiCl_4$ with $AlR_{3-y}(X)_y$ to give a brown precipitate active for olefin polymerization (1). Second generation catalysts have been made by depositing the active species on a high surface area support, such as silica, to increase polymerization activity and provide more desirable catalyst and polymer morphologies (2,3).

Even though extensive research has been done on unsupported and supported Ziegler-Natta catalysts for the polymerization of alpha olefins (4), many fundamental questions remain. For this reason a detailed study of silica supported titanium based Ziegler-Natta catalysts has been undertaken. The long term goal of this work is to

New Advances in Polyolefins, Edited by
T.C. Chung, Plenum Press, New York, 1993

correlate molecular level structure with catalytic properties. Therefore, the catalysts which are studied in this report are not as complex as commercial catalyst systems, to facilitate meaningful characterization.

To make catalyst structure/property correlations on silica supported catalysts, it is necessary to synthesize catalysts of discrete, determinable and variable molecular structures. The approach taken in this work is to modify the silica support with either thermal or chemical treatments prior to catalyst synthesis. The catalyst itself, in turn, is synthesized and analyzed as a chemically modified silica surface. Previous work on reactions of $TiCl_4$ with pre-modified silica gels describes the process in detail (5). Briefly, when $TiCl_4$ is reacted with a 600°C thermally pre-treated silica gel, a titanium species with one bond to the surface results:

$$Si(s)\text{-}OH + TiCl_4 \text{----}> Si(s)\text{-}O\text{-}Ti\text{-}Cl_3 + HCl \qquad \text{(Eq. 1)}$$

Alternatively, when $TiCl_4$ is reacted with a silica gel surface which has been previously silylated with hexamethyldisilazane (HMDS), a surface titanium species with two bonds (i.e. bridged) to the surface is obtained:

$$
\begin{array}{ll}
Si(s)\text{-}OH & Si(s)\text{-}O \\
\quad + TiCl_4 \text{----}> & \qquad TiCl_2 + 2HCl \\
Si(s)\text{-}OH & Si(s)\text{-}O
\end{array}
\qquad \text{(Eq. 2)}
$$

When $TiCl_4$ is reacted with unmodified silica gel, a mixture of singly and bridged bonded species result. These $TiCl_4$ reacted silica gels serve as model catalyst precursors for olefin polymerization catalysts.

In this report, studies are described concerning the reactions of silica gels with $TiCl_4$ and $Al(Et)_3$. Diffuse reflectance infrared Fourier Transform spectroscopy (DRIFTS) is used for much of the catalyst characterization. Ethylene homopolymerizations and ethylene/1-butene copolymerizations with $Al(Et)_3$ activated catalysts are also described. This is a first, tentative step towards the correlation of molecular level surface structure of silica supported Ziegler-Natta catalysts with catalytic properties.

EXPERIMENTAL

Reagents

Davison 948 silica gel was used as the catalyst support in all cases. This silica gel was prepared either by fluidization at 200°C in nitrogen for 4h, fluidization at 600°C in nitrogen for 4h, or reaction with HMDS, as described by Hoff and Pullukat (6), followed by fluidization in nitrogen at 200°C for 4h. Titanium tetrachloride and triethylaluminum (Aldrich, >99%) were stored under argon and used as received. Polymerization grade ethylene (Phillips) was used as received. Hexane (Fisher) was pre-dried with CaH_2 for a minimum of 24h prior to use. High purity argon (Matheson) was used as received. Butene-1 was treated with 13x molecular sieves prior to use.

Synthesis

Silica gel supports and catalysts were stored in a dry box (Vacuum Atmospheres Corp.). Solid reagents and catalysts were weighed in the dry box using pre-dried glassware subsequently sealed with rubber septa. Solid and liquid sample transfers were done using standard cannulae techniques. $TiCl_4$ reactions were done using neat reagent under argon for 1h at room temperature. The slurry was filtered, washed three times with hexane to remove physisorbed species, and evacuated overnight at room temperature to remove physisorbed species and residual solvent.

Polymerization Reactions

Polymerizations were done at atmospheric pressure in a glass vacuum line. Reactions were carried out for 3h at 27-31°C. Typically, 0.2-0.3g of $TiCl_4$ reacted silica gel was slurried with 100 ml of hexane. After transfer of this slurry to the polymerization vessel, triethylaluminum was added (Al/Ti = 3/1). The solid immediately turned dark upon addition of $Al(Et)_3$. The slurry was stirred vigorously with a magnetic stir bar prior to and during addition of ethylene or ethylene/1-butene mixtures. Rates of monomer addition were set by separate mass flow controllers (Sierra) at a rate of 50 cc/min.

Characterization

FTIR spectra were obtained using a Nicolet 60SX FTIR spectrometer purged with dry nitrogen and equipped with a liquid nitrogen cooled, narrow band mercury-cadmium-telluride detector. Spectra were acquired at 4 cm^{-1} nominal resolution by coaddition of 256 scans. Spectral subtractions were carried out using standard Nicolet software.

Catalyst spectra were obtained using the diffuse reflectance sampling technique. All samples were prepared in a glove box. A 10% w/w dispersion of sample was mixed in predried, ground KCl . Spectra were obtained using a commercially available diffuse reflectance accessory (Harrick Scientific, DRA-2CN) equipped with an inert atmosphere attachment (HVC-DRP). The accessory was modified as previously described (7). Infrared spectra of polyethylenes were obtained by hot pressing films and using conventional transmission techniques.

Elemental analyses were done using a Siemens SRS 300 wavelength disperisve x-ray fluorescence spectrometer. 0.3 g of sample was mixed with 8.7 g of $Li_2B_4O_7$ in a dry box. The sample was transferred to an automatic fluxer crucible and transformed into a glass disk. The weight percentage of titanium was obtained by measurement of the peak intensity at a two-theta angle of 86.187°. X-ray photoelectron spectroscopy was done with an AEI ES-100 spectrometer. All samples were handled in an inert atmosphere box where they were mounted in the analysis feedthrough chamber of the spectrometer. The spectrometer is equipped with a Henke x-ray source and a double focussing hemispherical analyzer with a retardation lens. Data collection and analysis is automated. A custom software package enabled spectral analysis and peak filling.

RESULTS AND DISCUSSION

In Figures 1-3, infrared spectra of chemically modified silica gels in the OH stretching region are shown. Spectra in Figures 1A, 2A, and 3A provide a comparison of silica surface silanol (Si-OH) groups on 200°C pretreated, 600°C pretreated, and HMDS reacted/200°C pretreated silica gels respectively. The 200°C thermally pretreated silica gel has three absorbances from non- or weakly hydrogen-bonded silanols (sharp peak at 3745 cm^{-1}), weakly hydrogen bonded vicinal silanols (3660 cm^{-1}), and strongly hydrogen bonded vicinal silanols (3540 cm^{-1}). A 600°C thermal pretreatment results in the loss of the latter two peaks from vicinal silanols (Figure 2A), while the non-hydrogen bonded silanols remain. Conversely, HMDS reaction of the silica surface results in the loss of non-hydrogen bonded silanols, while weakly and strongly hydrogen bonded vicinal silanols remain.

Vicinal silanols are condensed by 600°C thermal treatment, and non-hydrogen bonded silanols are reacted by HMDS. This implies that the 200°C, 600°C, and HMDS pretreated silica gels form an ideal triad for studying the role of various surface hydroxyl groups on reaction chemistry. Do these broad generalizations of complex surface chemistry approximate reality? If the HMDS and thermal reactions are as clean as implied (and the FTIR data suggest), then the wt % of titanium on 200°C silica gel should approximately equal the sum of wt % Ti on HMDS silica plus the wt% Ti on 600°C silica. Data in Table 1 shows that this is the case.

The chemical and thermal pretreatments are thus complementary approximately in terms of eliminating the different surface groups.

Table 1 - Weight % Titanium of TiCl$_4$ reacted silica gels.

Wt. % Ti 200°C SiO$_2$	Wt. % Ti 600°C SiO$_2$	Wt. % Ti HMDS SiO$_2$	Wt. % Ti 600°C + HMDS SiO$_2$
5.18	3.33	1.96	5.29

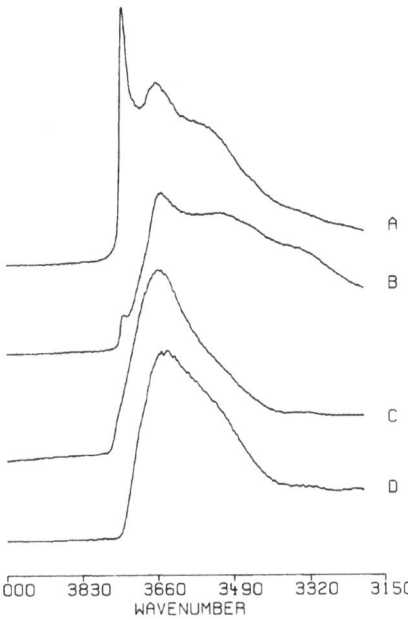

Figure 1 - Infrared spectra of 200·C - petreated silica gel. A) 200·C silica gel, B) after TiCl$_4$ reaction, C) after Al(Et)$_3$ reaction, D) TiCl$_4$ reaction followed by Al(Et)$_3$ reaction.

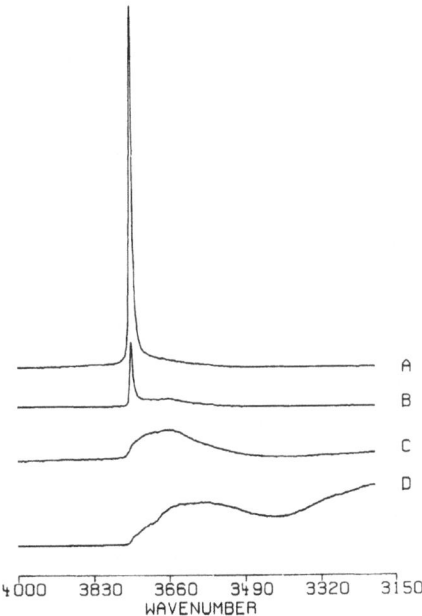

Figure 2 - Infrared spectra of 600·C - pretreated silica gel. A) 600·C silica gel, B) after TiCl$_4$ reaction, C) after Al(Et)$_3$, D) TiCl$_4$ reaction followed by Al(Et)$_3$ reaction.

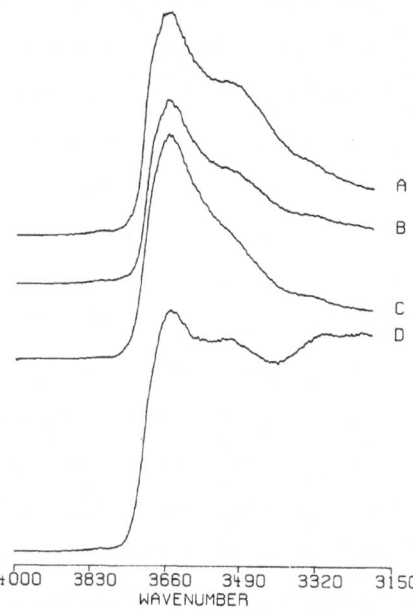

Figure 3 - Infrared spectra of HMDS - modified silica gel. A) HMDS silica gel, B) after TiCl$_4$ reaction, C) after Al(Et)$_3$ reaction, D) TiCl$_4$ reaction followed by Al(Et)$_3$ reaction.

Results in Figure 1 show changes in the infrared spectrum of 200°C thermally pretreated silica gel after reaction either with TiCl$_4$, Al(Et)$_3$, or TiCl$_4$ reaction followed by Al(Et)$_3$. These spectra have been normalized to the silica band at 1860 cm^{-1} so that spectral intensities can be compared. Comparison of Figures 1B and 1C indicates that TiCl$_4$ reacts extensively but not completely with non-hydrogen bonded silanols, whereas Al(Et)$_3$ reacts exhaustively with these groups. These spectra also show that the Al(Et)$_3$ reacted silica gel has lost the 3540 cm^{-1} band from strongly hydrogen bonded vicinal silanols, whereas the TiCl$_4$ reacted silica gel has considerable absorbance remaining in this region. The TiCl$_4$ reacted silica gel (Figure 1B) was subsequently reacted with Al(Et)$_3$ (Al/Ti ratio = 3/1). The infrared spectrum of the resulting brown solid is shown in Figure 1D. After reaction with TiCl$_4$, both strongly hydrogen-bonded and residual non-hydrogen-bonded silanols reacted with Al(Et)$_3$. These results indicate that Al(Et)$_3$ reacts more extensively with surface silanols than TiCl$_4$, suggesting that some aluminum alkyl probably reacts with residual surface silanols on the catalyst support.

Normalized spectra in Figure 2 show the infrared spectral changes of 600°C pretreated silica gel after various surface modification reactions. Quantitative analysis of the non-hydrogen-bonded silanol band before and after TiCl$_4$ reaction (Figure 2A,B) indicates that approximately 80% of non-hydrogen bonded silanols are reacted with TiCl$_4$. These results are in agreement with results shown in Figure 1, suggesting that TiCl$_4$ does not completely react with non-hydrogen-bonded silanols on silica gel.

A number of workers have shown that with fumed silicas such as Cab-O-Sil, $TiCl_4$ reactions result in complete loss of non-hydrogen bonded silanols (8-10). This work, and that reported by Ellestad et al. (11), were done with silica gels. Although Ellestad does not report quantitative data, his spectra show residual non-hydrogen bonded silanols remaining after $TiCl_4$ reaction. Data in this report suggests there is approximately 20% of non-hydrogen-bonded silanols on silica gel surfaces which are apparently unreactive to $TiCl_4$ under these conditions. Reaction of $Al(Et)_3$ with the 600°C pretreated silica gel results in the complete loss of non-hydrogen-bonded silanols. This result is also similar to the $Al(Et)_3$ reaction with 200°C silica gel (Figure 1C) where all non-hydrogen-bonded silanols are consumed. The regeneration of a small amount of hydrogen bonded hydroxyl groups could be derived from partially hydrolyzed $Al(Et)_3$ starting material. Figure 2D is a spectrum of the $TiCl_4$ reacted silica gel (Figure 2B) after reaction with $Al(Et)_3$ (Al to Ti ratio = 3/1). Some $Al(Et)_3$ has reacted directly with residual non-hydrogen bonded silanol groups and once again some hydrogen bonded hydroxyls are seen with $Al(Et)_3$.

Figure 3 illustrates spectra for the same reactions as Figures 1 and 2 for HMDS-modified silica gel. These spectra are also normalized so that intensity differences can be compared. The $TiCl_4$ reacted sample (B) shows a slight attenuation in both the 3660 and 3540 cm^{-1} bands compared to the HMDS silica gel (A). The $Al(Et)_3$ reacted sample (C) shows a more pronounced attenuation of the 3540 cm^{-1} band indicating more extensive $Al(Et)_3$ reaction with these groups compared to $TiCl_4$. The spectrum in Figure 3D of $TiCl_4$ reaction (Figure 3B) followed by $Al(Et)_3$ treatment shows a more pronounced baseline change similar to spectra in Figures 1D and 2D.

Traditionally, infrared spectroscopy of silica gels has been limited to the regions between 4000 and 1300 cm^{-1} due to the extremely intense silica matrix (Si-O-Si) absorptions. Application of the surface selective diffuse reflectance sampling technique makes it possible to obtain spectral information below 1300 cm^{-1}, where silicon-oxygen-metal absorptions should occur. Spectra of bare silica gel before and after reaction with $TiCl_4$, in the region below 1300 cm^{-1}, are shown in Figure 4A. Although minor differences can be seen, these differences are more apparent after spectral subtraction (Figure 4B). The spectrum in Figure 4B consists of bands at 990 and 920 cm^{-1}. The 990 cm^{-1} absorption has been previously assigned to singly bonded groups, the 920 cm^{-1} absorption from doubly or bridged bonded groups (5,12-14).

In Figure 5, difference spectra of $TiCl_4$/silica minus silica, $TiCl_4$/HMDS silica minus HMDS silica, and $TiCl_4$/600°C silica minus 600°C silica are shown. Subtractions were carried out by scaling to the silica gel absorbance at approximately 800 cm^{-1}. The difference spectrum of $TiCl_4$/silica minus silica contains bands at 990 and 920 cm^{-1} from singly and bridged bonded groups, respectively. The difference spectrum of $TiCl_4$ reacted HMDS modified silica gel has a band at 920 cm^{-1} which predominates from bridged bonded groups. Conversely, the difference spectrum of $TiCl_4$ reacted/600°C pretreated silica gel has a band at 990 cm^{-1} which predominates from singly bonded groups. These results show that predominantly singly bonded, bridged bonded, or mixed titanium surface species can be synthesized. These are

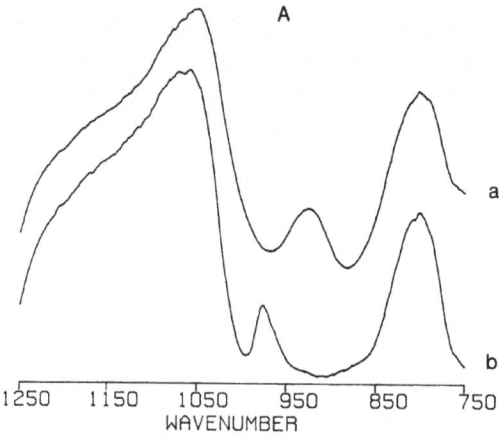

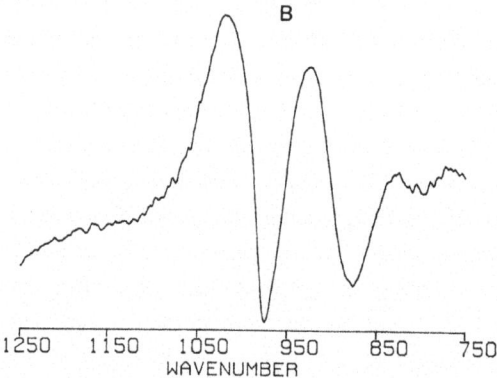

Figure 4 - (A) Infrared spectra of unmodified silica gel a) after TiCl$_4$ reaction and b) before TiCl$_4$ reaction. B) Result of spectral subtraction.

excellent model silica supported olefin polymerization pre-catalysts.

To correlate molecular level surface structure with catalytic properties, it is not only necessary to understand the structure of the modified silicas after reaction with TiCl$_4$, but to have some assurance that reaction of Al(Et)$_3$ with TiCl$_4$ silica does not result in the breaking of bonds which anchor the titanium to the surface. Some workers have suggested that reaction of aluminum alkyls with silica can result in the breaking of bonds (12,15), and the formation of TiCl$_3$ crystallites on the surface (16). Ellestad (17) has concluded that chloroaluminum alkyls can "extract" surface bonded titanium, whereas aluminum alkyls like Al(Et)$_3$ do not. To test whether Al(Et)$_3$ breaks surface Si-O-Ti bonds in this system, a series of spectral subtractions were done. Results of spectral subtraction for surface modified 600°C silica gel are shown

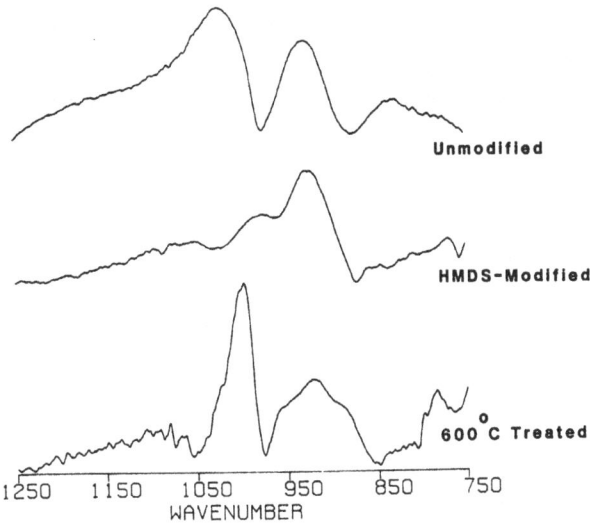

Figure 5 - Subtracted spectra of TiCl$_4$ reacted silica gels.

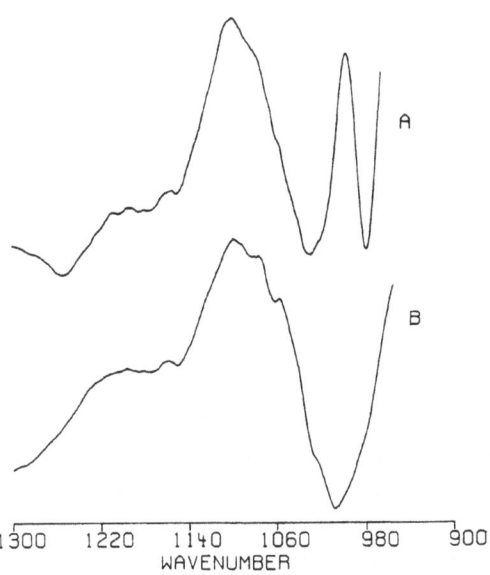

Figure 6 - Infrared results from spectral subtraction. A) Al(Et)$_3$/TiCl$_4$/ 600·C SiO$_2$ - 600·C SiO$_2$, B) Al(Et)$_3$/TiCl$_4$/600·C SiO$_2$ - TiCl$_4$/600·C SiO$_2$

in Figure 6. The top spectrum, of 600°C silica gel subtracted from the sample after reaction with $TiCl_4$ and $Al(Et)_3$, contains bands at 1090 and 990 cm^{-1} assigned to singly bonded Si-O-Al (tentatively) and Si-O-Ti species, respectively. The Si-O-Al bonds are not surprising since results in Figure 2B show that reaction with $TiCl_4$ leaves residual hydroxyl groups which are susceptible to attack by $Al(Et)_3$ forming Si-O-Al bonds. The existence of a strong 990 cm^{-1} band indicates that many Si-O-Ti species remain. Figure 6B is the spectrum which results from subtraction of $TiCl_4$ reacted silica from $TiCl_4/Al(Et)_3$ silica gel. If $Al(Et)_3$ cleaves Si-O-Ti bonds, a negative going band at 990 cm^{-1} would result. The lack of a significant negative going band at 990 cm^{-1} indicates that few if any Si-O-Ti bonds are cleaved under these conditions. Results are similar for the 200°C and HMDS silica gel series (not shown).

The activity for ethylene homopolymerization was tested for the catalyst supported on 600°C silica gel and the catalyst supported on HMDS-modified silica gel. This is a comparison of polymerization activity for singly bonded versus bridged bonded titanium surface species. For comparison, a 600°C silica gel first reacted with dibutylmagnesium then $TiCl_4$ was also synthesized and used for polymerization. Results are shown in Table 2.

Table 2 - Ethylene homopolymerization data of model catalysts.

Catalyst	Activity (g PE/g Ti/h)	Activity (g PE/g Cat/h)
HMDS SiO$_2$ TiCl$_4$/TEAL	50	0.7
600°C SiO$_2$ TiCl$_4$/TEAL	170	3.1
600°C SiO$_2$ DBM/TiCl$_4$/TEAL	-	21

It should be noted that the polymerizations were carried out at atmospheric pressure so these numbers do not correspond to the high pressure conditions more generally used. The data indicates that singly bonded surface titanium species are approximately 3.5 times more active towards the polymerization of ethylene compared to the bridged bonded groups. In both cases the polyethylene produced is very high molecular weight (no H$_2$ present in the reactor) and linear as evidenced by a lack of a 1379 cm^{-1} absorption from methyl groups in the infrared spectrum (Figure 7). The data in Table 1 also shows the large affect of magnesium on the silica supported catalysts. The presence of magnesium apparently can give nearly an order of magnitude enhancement in activity over catalysts without magnesium present. This is in agreement with work reported by Hsieh et. al. (18).

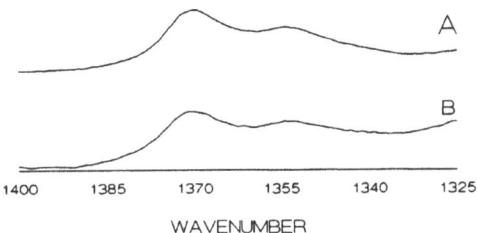

Figure 7 - Infrared spectra (in absorbance) of polyethylene from ethylene homo-polymerization: A) Al(Et)$_3$/TiCl$_4$/600°C SiO$_2$ catalyst, B) Al(Et)$_3$/TiCl$_4$/HMDS SiO$_2$ catalyst.

Table 3 - Data from x-ray photoelectron spectroscopy of model catalysts.

Catalyst	Al/Ti Ratio	Ti 2p 3/2 Binding Energy (eV)
200°C SiO$_2$	0	458.1
200°C SiO$_2$	3	458.1
HMDS SiO$_2$	0	458.0
HMDS SiO$_2$	3	457.9
600°C SiO$_2$	0	457.5
600°C SiO$_2$	3	457.5

The enhanced polymerization activity for singly bonded relative to bridged bonded surface titanium groups can be rationalized with electronic and/or steric arguments. To probe the electronic environment of surface bonded titanium species, x-ray photoelectron spectroscopy (XPS) was used. Results are shown in Table 3. The XPS data shows that singly bonded titanium surface species have a lower binding energy than the bridged species. This indicates that the singly bonded titanium groups are more electron rich, which could facilitate reduction by Al(Et)$_3$. Chien has indirectly shown (19) that singly bonded groups are more easily reduced than bridged titanium surface species which could explain the observed difference in polymerization activity. Reduction of some titanium surface species has occurred on all catalysts after reaction with Al(Et)$_3$ evidenced by the dark brown color of the materials, and the fact that they are polymerization active. Strangely, XPS detects no shift in titanium binding energy after reaction with Al(Et)$_3$ (Table 3). This may be because only a small percentage of these species are reduced. From a steric viewpoint, it has been suggested (19) that doubly attached species are immobile on the surface, whereas

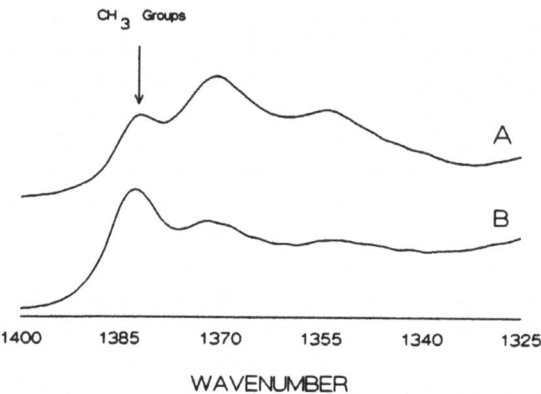

CH₃ Groups

| 1400 | 1385 | 1370 | 1355 | 1340 | 1325 |

WAVENUMBER

Figure 8 - Infrared spectra (in absorbance) of polyethylene from ethylene/
1-butene copolymerization: A) Al(Et)$_3$/TiCl$_4$/600·C SiO$_2$ catalyst, B) Al(Et)$_3$/
TiCl$_4$/HMDS SiO$_2$ catalyst.

singly bonded groups can rotate. This steric factor may hinder ethylene insertion at
the active site negatively affecting activity.

Some copolymerization experiments with mixtures of ethylene and 1-butene
were also performed. The catalyst with singly bonded surface titanium species gave
a polymer whose infrared spectrum is shown in Figure 8A. The presence of a 1379
cm^{-1} band indicates that this has a small number of ethyl branches. Further indication
that this polymer contains some short chain branches comes from the DSC melting
point of the copolymer (127°C compared to the two homopolymers at 133°C). Under
the same reaction conditions, copolymerization of ethylene and 1-butene with the
bridged bonded catalyst gave a polymer with infrared spectrum shown in Figure 8B.
The spectrum shows a much more intense absorption at 1379 cm^{-1} indicating
significantly more 1-butene insertion. The DSC melting point of this polymer was
125°C. Apparently, the bridged bonded catalyst has a greater propensity to
copolymerize 1-butene than the singly bonded catalyst. Once again, steric and/or
electronic arguments must be invoked to explain these results. Soga et. al. (20) and
others (21) have suggested that Ti(II) will selectively polymerize ethylene in an
ethylene/propylene mixture, whereas Ti(III) is active towards ethylene/propylene
copolymerization. These results may be consistent with what is seen here, since Chien
concluded (19) that singly bonded catalysts give predominantly Ti(II), whereas bridged
bonded catalysts give predominantly Ti(III) after reaction with aluminum alkyl.
Further work in characterizing the activated catalysts to separate various factors
controlling these polymerizations will be necessary to more fully explain the
polymerization data.

ACKNOWLEDGMENTS

I wish to thank Ian Peat, Jawad Murib, Steve Landau and many others for their help and support. I also thank Quantum Chemical Corporation for supporting and allowing publication of this work.

REFERENCES

1. G. Natta, P. Pino, P. Corradini, F. Danusso, E. Mantica, G. Hazzanti, and G. Moraglio, *J. Am. Chem. Soc.* 77:1708 (1955).

2. A. Munoz-Escalona, J.G. Hernandez, and J.A. Gallardo, *J. Appl. Polym. Sci.* 29:1187 (1984).

3. Belg. 839,870 (1976), Union Carbide Corp.

4. J. Boor, Jr. "Ziegler-Natta Catalysts and Polymerizations," Academic Press, New York (1979).

5. J.P. Blitz, *Colloids and Surfaces* 63:11 (1992).

6. U.S. 4,530,912 (1985), Chemplex Corp.

7. R.S.S. Murthy, J.P. Blitz, and D.E. Leyden, *Anal. Chem.* 58: 3167 (1986).

8. I. Bakardjiev, M. Majdraganova, and G. Blizanokov, *J. Non-Cryst. Solids* 20:349 (1976).

9. C.G. Armistead, A.J. Tyler, F.H. Hambleton, S.A. Mitchell, and J.A. Hockey, *J. Phys. Chem.* 73:3947 (1969).

10. M.L. Hair, and W. Hertl, *J. Phys. Chem.* 77:2070 (1973).

11. O.H. Ellestad, and U. Blindheim, *J. Mol. Catal.* 33:275 (1985).

12. J.B. Kinney, and R.H. Staley, *J. Phys. Chem.* 87:3735 (1983).

13. B.A. Morrow, C.P. Tripp, and R.A. McFarlane, *J. Chem. Soc, Chem. Commun.* 1282 (1984).

14.   A.V. Kiselev, A.I. Kuznetsov, V.I. Lynin, Yu. I. Rostorguev, B. I. Shalomov, and K. L. Shchepalin, *Kolloidn. Zh.* 42:969 (1980).

15.   J. Murray, M.J. Sharp, and J.A. Hockey, *J. Catal.* 18:52 (1970).

16.   J. Kratochvila, T. Shiomo, and K. Soga, *Makromol. Chem., Rapid Commun.* 11:541 (1990).

17.   O.H. Ellestad, *J. Mol. Catal.* 33:289 (1985).

18.   H.L. Hsieh, M.P. McDaniel, J.L. Martin, P.D. Smith, and D.R. Fatey *in*: "Advances in Polyolefins", R.B. Seymour, T. Cheng, Eds. Plenum Press, New York (1987).

19.   J.C.W. Chien, *J. Catal.* 23:71 (1971).

20.   K. Soga, K. Izumi, Y. Sano, and M. Terano, *Shokubai* 23:124 (1981).

21.   N. Kashiwa, and J. Yoshitake, *Makromol. Chem.* 185:1133 (1984).

STATISTICAL PROPAGATION MODELS FOR ZIEGLER-NATTA POLYMERIZATION

H. N. Cheng

Hercules Incorporated, Research Center
500 Hercules Road
Wilmington, Delaware 19808

ABSTRACT

Statistical probability models are widely used as a convenient theoretical framework for the characterization of polyolefins. Other than the commonly known enantiomorphic-site, Bernoullian, and Markovian models, many additional models have been devised in the last few years. This work surveys the various models proposed for the description of Ziegler-Natta polymerization. The models may involve propagations that are controlled by monomer units at the propagating chain ends or by the catalytic sites. The models may involve multiple active sites. They may involve an active site that switches back and forth between two states. Finally, the Markovian or enantiomorphic-site probabilities may be perturbed, leading to compositional heterogeneity. These models and the analytical methodologies involved are applied to published NMR data to test for the presence of dual catalytic-site/chain-end control and tacticity heterogeneity (distribution) in polypropylene made with Group IV metallocene catalysts, and to check for blocky microstructure in a polymer where the catalyst may exhibit inversion of stereocontrol.

INTRODUCTION

Statistical propagation models have been used extensively in the NMR characterization of polymers[1-4]. They provide fundamental understanding of the polymerization processes and serve as the theoretical framework whereby NMR, molecular weight, fractionation, and monomer feed/composition data can be analyzed. In many cases the models can assist in NMR spectral assignments and help to interpret the spectral intensities. Suitably applied, they may also provide information on the mechanisms of initiation and propagation.

Over the years, many statistical models have been used[1-16] for polyolefins. The combination of NMR and statistical models is particularly useful in characterizing polymer microstructures and elucidating polymerization mechanisms. Recently, a number of statistical propagation models have been formulated[17-25], many of them having direct applications to polyolefins. In this work an attempt has been made to survey the relevant models, to show their interrelationships, and to indicate the potential areas of application.

New Advances in Polyolefins, Edited by
T.C. Chung, Plenum Press, New York, 1993

ONE-STATE MODELS

In the simplest statistical models, one assumes that the polymer is made homogeneously through one reaction mechanism conforming with one statistical model. Two categories of models are available: chain-end control and catalytic-site control. In the chain-end control models, the probabilities of monomer insertion and its stereo- and regio-chemistry are determined by the identity and the configuration of the monomer residue(s) at the propagating chain end. In the catalytic-site control models, the probability of monomer insertion is dictated by the catalytic site only. Separate considerations need to be made for copolymer sequence and homopolymer tacticity.

Copolymer Sequence. For a copolymer consisting of monomers M_1 and M_2 where the monomer insertion is catalytic-site controlled, we can write the propagation reaction as follows:

$$-* + M_1 \xrightarrow{k_1} - M_1* \qquad (1a)$$

$$-* + M_2 \xrightarrow{k_2} - M_2* \qquad (1b)$$

where - represents the polymer chain, * represents the metal in the catalytic site, and k_1 and k_2 are the reactivities of the comonomer in the insertion reaction (or alternatively the selectivity of the catalytic site for the comonomer). The reactivity ratio r can be defined: $r = k_1/k_2$. The reaction probabilities for M_1 and M_2 (P_1 and P_2, respectively) are:

$$P_1 = \frac{rx}{1 + rx}, \qquad P_2 = \frac{1}{1 + rx}, \qquad (2)$$

where $x = (M_1)/(M_2)$, and (M_1) and (M_2) are the feed concentrations of M_1 and M_2. Thus, the site-controlled propagation of copolymer is governed by the preference of the site for the comonomers (as measured by the reactivity ratio r) and by the availability of the comonomers (as measured by the feed concentration ratio x). Note that these expressions are the same as those given for the Bernoullian (B) statistics.

When the chain ends also participate in controlling the propagation, we need to use the Alfrey-Mayo treatment[26,27]:

$$- M_1* + M_1 \xrightarrow{k_{11}^0} - M_1 M_1*$$

$$- M_1* + M_2 \xrightarrow{k_{12}^0} - M_1 M_2* \qquad (3)$$

$$- M_2* + M_1 \xrightarrow{k_{21}^0} - M_2 M_1*$$

$$- M_2* + M_2 \xrightarrow{k_{22}^0} - M_2 M_2*$$

The following well known equations can be derived for the reaction probabilities:

$$P_{12} = \frac{1}{1+r_1 x}, \qquad P_{21} = \frac{1}{1+r_2/x}, \qquad (4)$$

where the reactivity ratios $r_1=k_{11}/k_{12}$, $r_2=k_{22}/k_{21}$. This represents the first-order Markovian (M1) statistical model where the insertion of the monomer depends on the last unit in the propagating chain.

As derived, the values of r_1 and r_2 can reflect *both* the reactivity of comonomers M_1 and M_2 as well as the selectivity of the catalytic site for the comonomer. Thus, the M1 model in its general form assumes that polymer propagation is both catalytic-site and chain-end controlled.

16

Homopolymer Tacticity. In tacticity studies, a distinction needs to be made between Bovey formalism and Price formalism. Bovey formalism[1] considers the relative configuration, whereas Price formalism[2] considers the absolute configuration. The relationship between these formalisms and their relevance to NMR analysis have been pointed out earlier[18].

In the Price formalism, we can treat the tacticity problem as a copolymerization of levo (l) and dextro (d) configurations.

$$-* + M_d \xrightarrow{k_d} - M_d* \qquad (5a)$$

$$-* + M_l \xrightarrow{k_l} - M_l* \qquad (5b)$$

Since the monomer M is the same, x=1, and the reaction probabilities (P_d and P_l) are:

$$P_d = \frac{1}{1 + r} , \qquad P_l = \frac{r}{1 + r} , \qquad (6)$$

where $r=k_d/k_l$. This is the Bernoullian model in the Price formalism. As indicated earlier[18], this is equivalent to the enantiomorphic-site (E) model[5,7,8] commonly used to describe isospecific polymerization.

When chain ends also participate in controlling the propagation, expressions equivalent to Equations 3 and 4 can be written. The reaction probabilities are:

$$P_{dl} = \frac{1}{1 + r_d} , \qquad P_{ld} = \frac{r_l}{1 + r_l} , \qquad (7)$$

where r_d and r_l are the "reactivity ratios". This is the first-order Markovian (M1) model in the Price formalism. Two parameters are needed to describe the statistics: P_{dl} and P_{ld}. As in the copolymer case, the Price M1 model assumes that the propagation is controlled by both catalytic site and chain end. The same expressions have been derived by Furukawa et al[6], Ewen[23], and Cheng[19]. It is also called the E-M1 model[19].

A special case of the Price M1 model needs to be considered. If $P_{ld} = P_{dl} =$ probability of racemic diad, and $1-P_{ld} = 1-P_{dl} =$ probability of meso diad, then we can relax the constraint of absolute configuration. Only relative configuration needs to be considered, and only one parameter (P_m or P_r) is needed. This is the Bernoullian model in the Bovey formalism and represents the pure chain-end control mechanism.

A summary of the various one-state models is given in Table 1. The tacticity in the various models can be readily calculated. For example, the overall tacticity in the Price M1 model are:

(m) = (dd) + (ll) = $P_{ld}(1-P_{dl}) + P_{dl}(1-P_{ld})$
(r) = (dl) + (ld) = $2P_{ld}P_{dl}$

Table 1. One-state models and control mechanisms

	pure site control	pure end control	both site & end control
homopolymer tacticity	E (= Price B)	Bovey B (= spec. case of Price M1 model)	Price M1 (general case) Price M2
copolymer sequence	B	M1	M1 M2

TWO-STATE MODELS

The next level of complexity occurs when two separate states are involved in polymerization. The two states may correspond to two separate active sites, or may correspond to one active site that switches back and forth between two states with different propagation statistics.

Two types of models can be distinguished[22]. In the *consecutive two-state* models, a catalytic site switches back and forth between two states even as it polymerizes. The result is a block copolymer consisting of blocks that conform to different statistical (E, B, or Markovian) models. Sometimes, during polymerization the reaction condition is changed (e.g., sudden changes in the pressure of the feed gas or in the temperature); in this case, the polymer made during the change will also conform to the consecutive model.

In the *concurrent two-state* models, two separate active sites are present, each site making its own polymer according to its propagation statistics. In effect, a blend of two polymers is formed. Another case of concurrent two-state model may arise if a catalytic site switches back and forth between two states, but the rate of switching is low relative to chain propagation and chain termination. As a result, separate chains are formed that can be attributed to each of the two states.

The two-state B/B models (both consecutive and concurrent) have been treated in the classic work of Coleman and Fox[15]. The application of the Coleman-Fox model to NMR data has been described by Frisch, et al.[16] The concurrent B/E model has been used by Zambelli, et al[9] and Chujo, et al[10] for the treatment of polypropylene tacticity. The concurrent B/B model has been applied to olefin copolymers by Ross[11] and others[12-14]. In general, the treatment of the two-state concurrent models is similar to the treatment of multistate models[14] (see below).

Recently, a general treatment of the two-state consecutive models has been formulated.[22] A structural approach (different from Coleman-Fox) has been used and applied to B/B, B/E, and E/E cases. These models have been used on the tacticity data of polypropylene produced by non-symmetric ansa-titanocene catalysts[28].

A different two-state model has been proposed by Ewen, et al[24] to rationalize the NMR data of syndiotactic polypropylene made with metallocene catalysts. Chain-migratory insertion mechanism[29,30] has been presumed. The model includes dual catalytic-site and chain-end control. Another two-state probability model was recently devised by Cheng and Kasehagen[25], attempting to explain the NMR data of polypropylene in general.

MULTISTATE MODELS

It is well known that Ziegler-Natta catalysts may contain more than two active sites. The NMR data for a polymer made with such a catalyst would have multiple components. A general methodology for the treatment of such data has been proposed previously[14]. The results are found to be most definitive when NMR data of polymer fractions are available. The methodology has been applied to copolymer sequence analysis of ethylene/propylene copolymer[14,31,32], propylene/1-butene copolymer[14], ethylene/1-butene copolymer[33,34], and ethylene/1-hexene copolymer[35], and the tacticity determination of polypropylene[14,36] and poly(1-butene)[14,31].

PERTURBED MODELS

Recently, perturbed Markovian[21,50] and enantiomorphic-site[37] models have been devised to account for compositional and tacticity heterogeneity observed for many vinyl

polymers. Equations for the Bernoullian model have been derived earlier by Ross[11].
Compositional Heterogeneity can have many sources, some of which are shown below:

homopolymer tacticity	copolymer sequence
statistical	statistical
chain end	conversion
multistate	multistate
process	process

Multistate heterogeneity can be treated with multistate models. The other three kinds of heterogeneities can be described by the perturbed models. Using the perturbed model, one can analyze[21,37,50] or simulate[38,39] the NMR data and the compositional distribution curves.

Another application of the perturbation model may arise if the stereocontrol of the active site is partial or imprecise. The resulting reaction probability would then have a range of values: $P_{ij} = <P_{ij}> \pm \varepsilon$, where P_{ij} and $<P_{ij}>$ are the actual and the average reaction probabilities, respectively, and ε is a measure of the spread. A similar situation may be encountered in copolymerization if the selectivity for the comonomers A and B involves uncertainty (e.g., the active site exhibiting a distribution of selectivities). Again, the perturbation model may be applicable.

A summary of the propagation models is given in Figure 1. Also shown are the approximate relationship between the various models.

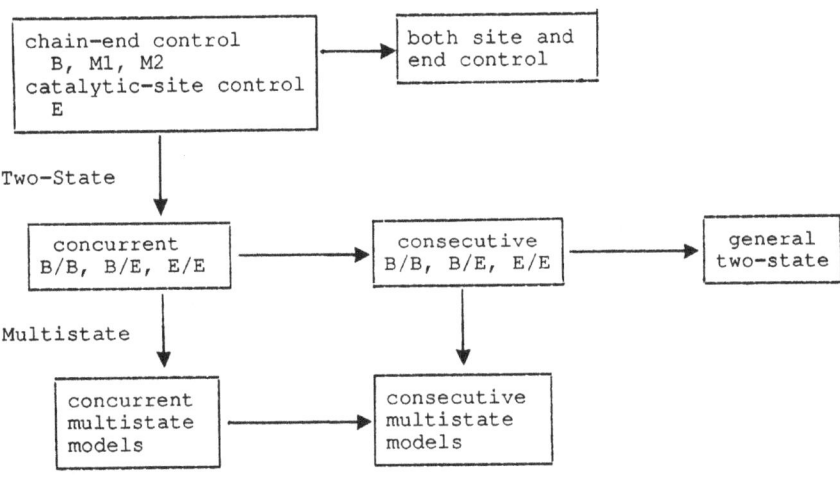

Figure 1. Summary of statistical propagation models

ANALYTICAL METHODOLOGY

In order to apply these various models to actual data, appropriate analytical methodologies are needed. As far as polymer microstructure and heterogeneity are concerned, two general approaches have been used. In the <u>analytical</u> (model-fitting) approach[14,17,36,40], the observed raw data (e.g., NMR spectrum or SEC chromatogram) are interpreted, and the processed information (e.g., sequence intensities, M_n, and M_w) is fitted to a reaction probability model through a suitable fitting algorithm:

$$\text{observed data} \longrightarrow \text{interpretation} \longrightarrow \text{fitting algorithm} \longrightarrow \text{probability model}$$

An alternative approach[17,19,38,39,41,42] is to start with the reaction probability model and simulate the polymerization process in a computer (e.g., through Monte Carlo simulation). The chains are then counted for specific information (e.g., polymer sequence, composition distribution, molecular weights, and even simulated NMR spectra). Because one starts with a model and builds polymer chains in a computer simulation process, this is called the <u>synthetic</u> (or computer simulation) approach.

$$\text{probability model} \longrightarrow \text{chain generation} \longrightarrow \text{bookkeeping} \longrightarrow \text{predicted results}$$

The advantages and disadvantages of both approaches have been previously discussed[17]. In general, the two approaches are complementary; when used together, they provide a good description of the polymer system in question.

In the past several years a number of computer programs have been developed in this laboratory to handle the various reaction probability models as pertaining to polyolefins. A listing of some of the programs are given in Table 2. Additional programs have been reported previously[17,50].

Table 2. Probability models and computer programs available

	Analytical	Synthetic
one-state models (chain-end and site control)	FITCO[18,40]	CALCO[41,42] PODIS[38] TADIS[39]
two-state models concurrent	MIXCO[14]	PODIS[38] TADIS[39]
consecutive	EXCO[22]	PODIS[38] TADIS[39]
multistate models concurrent	MIXCO[14]	PODIS[38] TADIS[39]
perturbed models	PERT[21,37]	PODIS[38] TADIS[39]

ILLUSTRATIVE EXAMPLES

A few examples using polypropylene are given here to illustrate the use of these approaches. Other examples have been reported previously[14,17-22,31-42].

Dual Catalytic-Site/Chain-End Control. An area of current research activity is the Group IV metallocene catalysts. A question may be asked whether these catalysts exhibit both catalytic-site and chain-end control. The possibility of such dual control in polyolefins has been discussed earlier[18,19,23,24].

As a test case, we can use the NMR data reported by Rieger et al[43]. An isotactic polypropylene sample was made using $Et[Ind]_2ZrCl_2$ and MAO at $0^\circ C$ and then fractionated. The pentad intensities of the fractions are given in Table 3. Analysis[18] can be made using first the enantiomorphic-site model, and then Price's M1 model which includes the effects of both the catalytic site and the chain end. The fitted results are also given in Table 3.

It appears that the predominant effect is the catalytic site. In the case of the ether-soluble, C_5-soluble, and C_6-soluble fractions, $P_{dd} \approx P_{1d}$; thus, only the zeroth order model (enantiomorphic site) is necessary. In the case of C_7 fractions, perhaps some chain-end control can be claimed ($P_{dd} > P_{1d}$). However, in this case an alternative interpretation of the data can be made via tacticity heterogeneity (see below).

Perturbed Models. It was recently shown[44] that the catalyst complexes involved in the metallocene/MAO system are fluxional due to torsional twisting of the ligands. There is also evidence that MAO is heterogeneous in composition[45-47], molecular weight[48], and may be engaged in different types of bonding with Zr[44]. Thus, in suitable cases, one might expect this type of catalysts to give polymers that exhibit some degree of tacticity heterogeneity.

In the case of the polymer fractions reported by Rieger. et al[43] there is strong evidence for tacticity heterogeneity. For example, the P_1 values in the five fractions range from 0.70 to 0.97 (Table 3). Furthermore, we can analyze the C7-soluble and C7-insoluble fractions by the perturbed E model. The calculated pentads and the reaction probabilities are given in Table 4. It appears that the perturbed model gives an improved fit over the pure E model. The mean deviation for the perturbed E model are comparable to the Price M1 model. Thus, in this case, the polymer made by the catalyst appears to have a distribution of tacticity.

It may be cautioned that the above results should not be generalized. This author has recently examined a large number of published data, and somewhat different results have been found for polymers made at different laboratories. Further details can be found elsewhere[37].

Multiple Components. It is well known that polymers made with conventional heterogeneous Ziegler-Natta catalysts have multiple components due to the presence of multiple active sites in the catalysts. In a recent analysis[36] of two polypropylene samples where detailed NMR and size exclusion chromatographic data on polymer fractions are available, at least four separate components (corresponding to four catalytic sites) have been identified.

Frequently such detailed molecular weight and NMR data on polymer fractions are not available. In these cases, it is common to use the two-state B/E model[9-10] as a simplification. In their work on two-state B/E models, Inoue, et al[49] have obtained heptane-soluble and heptane-insoluble fractions of a polypropylene sample made from a heterogeneous Ziegler-Natta catalyst. The tacticity pentad intensities have been reproduced in Table 5 (columns 2 and 5).

Table 3. Observed pentad data of polymer fractions reported by Rieger, et al[43], and analysis by the enantiomorphic-site and the Price M1 model[a]

frac.[b]		pentad intensities									MD[c]	E-model	Price M1 model	
		mmmm	mmmr	rmmr	mmrr	xmrx	rmrm	rrrr	rrrm	mrrm		P_l	P_{dd}	P_{ld}
EE-sol.	obsd.	25.8	15.9	5.0	14.9	12.9	5.2	5.2	7.4	7.8				
	E-model	25.8	16.5	3.3	16.5	13.1	6.6	3.3	6.6	8.3	0.97	0.762		
	Price M1	25.8	16.5	3.3	16.7	12.9	6.4	3.4	6.6	8.4	0.95		0.762	0.770
C5-sol.	obsd.	37.9	17.4	5.8	13.0	6.7	4.2	2.5	6.7	5.9				
	E-model	37.9	16.4	2.1	16.4	8.5	4.2	2.1	4.2	8.2	1.68	0.823		
	Price M1	37.9	16.4	2.1	16.4	8.4	4.2	2.1	4.2	8.2	1.68		0.823	0.826
C6-sol.	obsd.	50.7	16.9	5.2	12.9	5.2	1.7	1.3	0.9	5.2				
	E-model	50.7	14.8	1.2	14.8	4.9	2.5	1.2	2.5	7.4	1.42	0.873		
	Price M1	50.7	14.8	1.3	14.5	5.2	2.7	1.2	2.4	7.3	1.37		0.873	0.860
C7-sol.	obsd.	82.7	4.4	2.7	4.9	1.1	0.7	0.5	1.2	1.6				
	E-model	82.7	6.4	0.1	6.4	0.5	0.3	0.1	0.3	3.2	1.11	0.963		
	Price M1	82.7	6.2	0.2	4.9	1.9	1.2	0.1	0.2	2.4	0.87		0.964	0.779
C7-ins.	obsd.	88.7	3.6	1.8	2.9	0.8	0.7	0.2	0.2	1.1				
	E-model	88.7	4.3	0.1	4.3	0.2	0.1	0.1	0.1	2.1	0.70	0.976		
	Price M1	88.7	4.1	0.2	2.9	1.6	0.9	0.0	0.1	1.4	0.42		0.978	0.706

[a] The Price M1 model (also called E-M1 model[19]) is characterized by the reaction probability P_{ij}, which is the probability of monomer j being enchained to a polymer terminating in monomer residue i. Note that Note that $P_{dd} + P_{dl} = 1$; and $P_{ld} + P_{ll} = 1$. Two parameters are needed; we use P_{dd} and P_{ld} here.

[b] EE = diethyl ether, C5 = n-pentane, C6 = n-hexane, C7 = n-heptane.

[c] mean deviation between the observed and the calculated pentad intensities.

Table 4. Observed pentad data of C7 fractions reported by Rieger, et al[43], and calculated intensities obtained through the perturbed enantiomorphic-site model[a]

frac.		mmmm	mmmr	rmmr	mmrr	xmrx	rmrm	rrrr	rrrm	mrrm	MD[b]	perturbed E model P_1	ε
					pentad intensities								
C7-sol.	obsd.	82.7	4.4	2.7	4.9	1.1	0.7	0.5	1.2	1.6	0.95	0.954	0.111
	calc.	82.7	4.9	0.5	4.9	2.0	1.0	0.5	1.0	2.5			
C7-ins.	obsd.	88.7	3.6	1.8	2.9	0.8	0.7	0.2	0.2	1.1	0.40	0.972	0.075
	calc.	88.7	3.6	0.2	3.6	0.9	0.5	0.2	0.5	1.8			

[a] Analyzed by program PERT; the parameters are P_1 and ε.

[b] Mean deviation between the observed and the calculated pentad intensities.

Table 5. Analysis of the pentad intensities of heptane-soluble and heptane-insoluble fractions of polypropylene through conventional and perturbed two-state E/B models.

pentad	heptane-soluble fraction			heptane-insoluble fraction		
	obsd[a] intensities	calc intensities E/B[b]	perturbed E/B[c]	obsd[a] intensities	calc intensities E/B[b]	perturbed E/B[c]
mmmm	39.3	39.3	39.3	82.7	82.8	82.8
mmmr	9.3	11.2	10.2	6.0	5.2	5.0
rmmr	3.0	1.3	1.9	1.2	0.2	0.4
mmrr	12.8	12.0	10.9	4.4	5.3	5.0
xmrx	7.4	8.4	9.5	1.3	1.1	1.6
rmrr	3.7	2.7	3.7	0.8	0.4	0.8
rrrr	12.7	12.7	12.7	1.1	1.5	1.1
rrrm	6.4	6.4	6.4	0.9	0.9	0.9
mrrm	5.3	6.0	5.4	1.5	2.7	2.5
E model: P_l		0.877	0.853		0.970	0.959
frac. wt.		0.757	0.781		0.965	0.988
B model: P_m		0.164	0.166		0.205	0.114
frac. wt.		0.243	0.219		0.035	0.012
ε		0	0.157		0	0.092
mean dev.		0.79	0.69		0.55	0.42

[a] observed intensities (in %) taken from ref. 49.
[b] analysis carried out with program FITCO/PPTAC (ref. 40).
[c] analysis carried out with program EPPTAC.

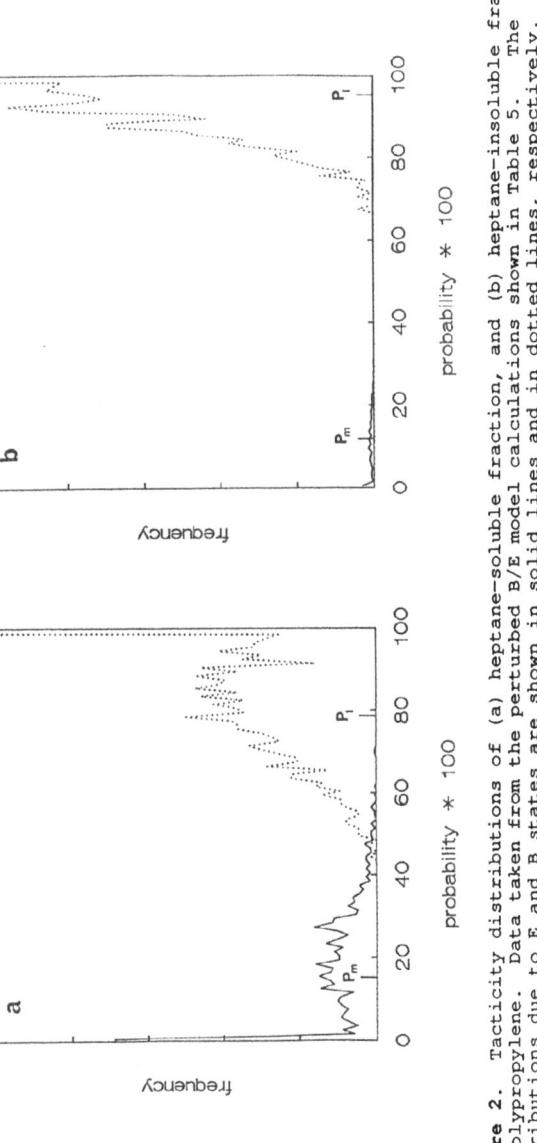

Figure 2. Tacticity distributions of (a) heptane-soluble fraction, and (b) heptane-insoluble fraction of polypropylene. Data taken from the perturbed B/E model calculations shown in Table 5. The distributions due to E and B states are shown in solid lines and in dotted lines, respectively.

Through the use of program FITCO[40], the data are first analyzed with two-state B/E model (Table 5, columns 3 and 6). As expected, the results are similar to those reported by Inoue, et al.[49] Next, the data are analyzed with the perturbed two-state B/E model using the perturbed model (program EPPTAC)[37]. The addition of the perturbation ε improves the fit (Table 5, columns 4 and 7). The mean deviations between observed and calculated pentad intensities decrease by roughly 10-20% in both cases. The results suggest that both fractions have some degrees of compositional heterogeneity ($\varepsilon = 0.157$ and 0.092).

In general, when multiple components are present in a sample, each component (having different reaction probabilities) may have a different average composition (average tacticity in this case). Thus, a multicomponent polymer usually has a multi-modal tacticity distribution curve. Unless one carries out a detailed fractionation of the sample, it is not easy to extract information from this compositionally heterogeneous polymer. One measure of the compositional spread and the shape of the tacticity distribution curve is the second moment about the mean[21,38,50]. Thus, the wider the spread of these components, the larger is the second moment.

It was previously shown[21] that the perturbation ε is proportional to the second moment of the compositional (tacticity) distribution curve. The large ε values in Table 5 obtained for the two polymer fractions suggest that the polymer probably contains more than the two E and B sites. A schematic diagram showing the tacticity distribution is given in Figure 2. Judging from the size of the ε values, one may guess that there may be at least two enantiomorphic sites, one site being more stereospecific ($P_1 > 0.90$), and the other less stereospecific ($P_1 < 0.90$). Likewise, there is at least one Bernoullian sites and maybe even two Bernoullian sites. For a detailed analysis of the various discrete catalytic sites, the polymers should be separated into more fractions, the NMR of the fractions obtained, and mixture analysis[14,31,32] (e.g., program MIXCO) applied.

Inversion of Stereocontrol. Busico, et al[51] recently reported an interesting study of the heterogeneous catalyst system $TiCl_4$/Na-octyl which produced polymers with unusually high molecular weights and high amounts of syndiotactic sequences. On the basis of their results, they concluded that a fraction of the active catalytic sites was able to invert their stereocontrol (from isospecific to mostly syndiotactic) during the growth of polymer chains. In view of their conclusion, it would be of interest to see if their NMR data conform to the *consecutive* two-state model.

Busico et al[51] have fractionated the polymer made with $TiCl_4$/Na-octyl with diethyl ether, hexane, and heptane, and reported the NMR pentad data (Table 6). They have made a careful analysis of their data through the *concurrent* two-state models (both B/E and E/E). They showed that their data are consistent with the concurrent two-state B/E model.

An analysis is carried out of Busico's data using the aforementioned methodologies. The data are first analyzed to see if the polymer contains multiple components. The pairwise analysis technique[14,31] is used, where the NMR data of two fractions are taken together and analyzed at once. The process is repeated for the other pairs. After initial calculations, it appears that a three-component (E/E/B) model can fit the entire NMR data shown in Table 6. The three components are characterized by the following reaction probabilities:

E_1-component:	$P_1 = 0.95$,
E_2-component:	$P_1 = 0.70 - 0.81$,
B-component:	$P_m = 0.12 - 0.15$.

The component weight factors and the approximate calculated pentad intensities are also given in Table 6. As noted earlier[32,36], the analysis merely indicates that the polymer contains *at least* three components; more components may actually be present, but

Table 6. Observed pentad intensities for a polymer made with TiCl$_4$/Na-octyl[51] and approximate calculated intensities through the concurrent three-component (E/E/B) model[a]

frac.[b]		pentad intensities									state 1		state 2		state 3	
		mmmm	mmmr	rrmr	mmrr	xmrx	rmrm	rrrr	rrrm	mrrm	P_1	w_1	P_1	w_2	P_m	w_3
C7-ins.	obsd.	63.0	7.4	2.1	7.4	5.3	1.1	6.3	3.2	4.2	0.95	0.76	0.81	0.12	0.15	0.13
	calc.	63.0	8.0	0.6	8.3	4.2	1.3	7.2	3.2	4.2						
C7-sol.	obsd.	42.0	8.6	2.0	10.0	8.3	3.0	15.1	6.0	4.2	0.96	0.40	0.78	0.35	0.13	0.25
C6-ins.	calc.	42.3	8.8	1.4	9.4	8.8	2.9	15.2	6.5	4.7						
C6-sol.	obsd.	24.9	6.4	2.8	7.8	11.6	2.8	31.2	9.9	2.4	0.95	0.25	0.73	0.29	0.13	0.46
EE-ins.	calc.	25.0	6.9	1.8	7.9	12.8	3.6	27.9	10.2	3.9						
EE-sol.	obsd.	12.3	10.0	3.7	12.3	18.5	5.9	17.9	12.7	6.6	0.94	0.00	0.70	0.73	0.14	0.27
	calc.	12.2	11.4	3.7	12.1	18.0	7.3	17.9	11.4	6.0						

a   The model assumes that each pentad intensity is the sum of three components: $I_{calc} = w_1 I_1 + w_2 I_2 + w_3 I_3$, where w_i is the component weight factor for a given fraction, and I_i is the theoretical intensities given by the E or B model (characterized by the probabilities, P_1 or P_m, respectively).

b   C7 = n-heptane, C6 = n-hexane, EE = diethyl ether.

Table 7. Testing of the NMR data of Busico's fractions[51] using two-state (E/B) models

frac.	concurrent (MIXCO)[a]					consecutive (EXCO)[b]				
	E-state		B-state			E-state		B-state		
	P_1	w_1	P_m	w_2	MD[c]	P_1	β	P_m	α	MD[c]
C7-ins.	0.941	0.855	0.150	0.145	0.71	0.941	93	0.150	21	0.85
						0.941	1060	0.150	180	0.71
						0.941	5130	0.150	870	0.71
C7-sol.	0.903	0.706	0.150	0.294	0.57	0.909	125	0.150	53	0.60
C6-ins.						0.902	1550	0.150	630	0.58
						0.903	5420	0.150	2260	0.57
C6-sol.	0.893	0.439	0.150	0.561	0.75	0.905	77	0.150	102	0.78
EE-ins.						0.897	240	0.150	310	0.76
						0.894	970	0.150	1250	0.75
EE-sol.	0.701	0.717	0.150	0.283	0.58	0.709	74	0.150	31	0.53
						0.703	330	0.150	130	0.57
						0.702	550	0.150	220	0.58

a   Analyzed by program MIXCO. Pentad intensities assumed to be sum of two states: $I_{obsd} = w_1 I_1 + w_2 I_2$, where w_i and I_i are defined as in Table 4.

b   Analyzed by program EXCO. α and β are the lengths of the blocks (in monomer units) obeying B and E models, respectively. Only selected iterations shown.

c   Mean deviation between the observed and the calculated pentad intensities.

information on them cannot be easily extracted from the NMR data alone.

It is of interest that all the fractions contain a Bernoullian component with $P_m \approx$ 0.15. The question is whether this Bernoullian component is a separate polymer present as a mixture (the concurrent case) or chemically attached to the enantiomorphic polymer chains (the consecutive case). The approach is to analyze the NMR results of each of the fractions by the consecutive two-state (E/B) model and compare the results with the concurrent two-state (E/B) model. To minimize the number of adjustable parameters, we shall keep $P_m = 0.15$ (invariant).

(It may be noted that in each fraction at least two E state are present. However, the probability expressions for the E states are exactly the same. For our purpose, the two E states can be approximated by an average E state. We can thus keep our analysis to the simpler, albeit slightly artificial, two-state case.)

The results of the calculations (from programs EXCO and MIXCO) are shown in Table 7. In the ether soluble fraction, a preference for the consecutive E/B model is indicated. The average block length for the E part (β) is 74; the average block length for the B part (α) is 31. In the other three fractions, the mean deviations between the observed and the calculated pentad intensities decrease steadily as block lengths increase. As a result, the NMR analysis alone cannot prove whether the consecutive model is valid. If it is, the block lengths (α and β) are larger than 1000. This is not unreasonable in view of the high molecular weights observed for these three fractions (degree of polymerization all larger than 8000).

On the basis of DSC and solubility behavior, Busico et al[51] have shown that the heptane-insoluble fraction is the product of inversion of stereocontrol. It appears that in the ether-soluble fraction the NMR data also provide evidence for this inversion mechanism.

Acknowledgments

The author is fortunate to have collaborated with a number of capable people over the years, viz., G. N. Babu, M. A. Bennett, J. C. W. Chien, J. A. Ewen, L. J. Kasehagen, R. A. Newmark, M. T. Roland, and S. B. Tam. This is Hercules Research Center Contribution Number 2162.

REFERENCES

1. Bovey, F. A. High Resolution NMR of Macromolecules; Academic Press: New York, 1969.
2. Price, F. P. In Markov Chains and Monte Carlo Calculations in Polymer Science; Lowry, G. G., Ed.; Marcel Dekker: New York, 1970, Chap. 7.
3. Koenig, J. L. Chemical Microstructure of Polymer Chains; Wiley: New York, 1980.
4. Cheng, H. N. In Modern Methods of Polymer Analysis; Barth, H. G.; Mays, J. W., Eds.; Wiley: New York, 1991, Chap. 11.
5. Sheldon, R. A.; Fueno, T.; Tsunetsuga, T.; Furukawa, J. J. Polym. Sci. 1965, B3, 23.
6. Furukawa, J. ACS Polymer Preprints 1967, 8(1), 39.
7. Doi, Y. Makromol. Chem., Rapid Comm. 1982, 3, 635.
8. Zambelli, A.; Locatelli, P.; Sacchi, M. C.; Tritto, I. Macromolecules 1982, 15, 831.
9. Zambelli, A.; Locatelli, P.; Provasoli, A.; Ferro, D. R. Macromolecules 1980, 13, 267.
10. Zhu, S.-N.; Yang, X.-Z.; Chujo, R. Polym. J.(Tokyo) 1983, 15, 859.

11. Ross, J. F. In Transition Metal Catalyzed Polymerizations, Alkenes and Dienes; R. P. Quirk, Ed.; Harwood Academic: New York, 1983, p. 799.

12. Cozewith, C. Macromolecules 1987, 20, 1237.

13. Floyd, S. J. Appl. Polym. Sci. 1987, 34, 2559.

14. Cheng, H. N. J. Appl. Polym. Sci. 1988, 35, 1639.

15. Coleman, B. D.; Fox, T. G. J. Chem. Phys. 1963, 38, 1065.

16. Frisch, H. L.; Mallows, C. L.; Heatley, F.; Bovey, F. A. Macromolecules 1968, 1, 533.

17. Cheng, H. N. J. Appl. Polym. Sci.: Appl. Polym. Symp. 1989, 43, 129, and references therein.

18. Cheng, H. N. J. Appl. Polym. Sci. 1988, 36, 229.

19. Cheng, H. N. In Transition Metal Catalyzed Polymerizations; Quirk, R. P., Ed.; Cambridge Univ. Press: Cambridge, 1988, p. 599.

20. Roland, M. T.; Cheng, H. N. Macromolecules 1991, 24, 2015.

21. Cheng, H. N. Macromolecules 1992, 25, 2351.

22. Cheng, H. N.; Babu, G. N.; Newmark, R. A.; Chien, J. C. W. Macromolecules 1992, 25, 6980.

23. Ewen, J. A. In Catalytic Polymerization of Olefins; Keii, T.; Soga, K., Eds.; Kodansha-Elsevier: Tokyo, 1986, p. 271.

24. Ewen, J. A.; Elder, M. J.; Jones, R. L.; Curtis, S.; Cheng, H. N. In Catalytic Olefin Polymerization; Keii, T.; Soga, K., Eds.; Kodansha: Tokyo, 1990, p.439.

25. Cheng, H. N.; Kasehagen, L. J., manuscript in preparation.

26. Alfrey, T.; Goldfinger, G. J. Chem. Phys. 1944, 12, 205.

27. Mayo, F. R.; Lewis, F. M. J. Am. Chem. Soc. 1944, 66, 1594.

28. Babu, G. N.; Newmark, R. A.; Cheng, H. N.; Llinas, G. H.; Chien, J. C. W. Macromolecules 1992, 25, 7400.

29. Cossee, P.; Arlman, E. J. J. Catal. 1964, 3, 80; also Rec. Trav. Chim. Pays-Bas 1966, 85, 1152.

30. Corradini, P.; Guerra, G. Prog. Polym. Sci. 1991, 16, 239, and references therein.

31. Cheng, H. N. ACS Symp. Ser. 1989, 404, 174.

32. Cheng, H. N.; Kakugo, M. Macromolecules 1991, 24, 1724.

33. Cheng, H. N. Polym. Bull. 1990, 23, 589.

34. Cheng, H. N. Macromolecules 1991, 24, 4813.

35. Cheng, H. N. Polym. Bull. 1991, 26, 325.

36. Cheng, H. N. Makromol. Chem., Theor. Simul. 1992, 1, 415.

37. Cheng, H. N. Makromol. Chem., Theor. Simul. 1993, 2, 561.

38. Cheng, H. N.; Tam, S. B.; Kasehagen, L. J. Macromolecules 1992, 25, 3779.

39. Cheng, H. N.; Kasehagen, L. J. Macromolecules, August 1993.

40. Cheng, H. N. J. Chem. Inf. Computer Sci. 1987, 27, 8.

41. Cheng, H. N.; Bennett, M. A. Anal. Chem. 1984, 56, 2320.

42. Cheng, H. N.; Bennett, M. A. Makromol. Chem. 1987, 188, 2665.

43. Rieger, B.; Mu, X.; Mallin, D. T.; Chien, J. C. W. Macromolecules 1990, 23, 3559.

44. Chien, J. C. W.; Sugimoto, R. J. Polym. Sci.: Polym. Chem. 1991, 29, 459.

45. Kaminsky, W.; Steiger, R. Polyhedron 1988, 7, 2375.

46. Resconi, L.; Bossi, S.; Abis, L. Macromolecules 1990, 23, 4489, and references therein.

47. Nekhaeva, L. A. J. Organomet. Chem. 1991, 406, 139, and references therein.

48. Cam. D.; Albizzati, E.; Cinquina, P. Makromol. Chem. 1990, 191, 1641.

49. Inoue, Y.; Itabashi, Y.; Chujo, R.; Doi, Y. Polymer 1984, 25, 1640.

50. Cheng, H. N. J. Appl. Polym. Sci.: Appl. Polym. Symp. 1992, 51, 21.

51. Busico, V.; Corradini, P.; DeBiasio, R.; Trifuoggi, M. Makromol. Chem. 1992, 193, 1765.

SUPPORTED CATALYSTS IN STIRRED BED GAS PHASE REACTORS

K.D. Hungenberg* and M. Kersting

Plastics Laboratory, BASF AG
D-6700 Ludwigshafen, Germany

INTRODUCTION

Polypropylene is one of the fastest growing plastic materials[1]. The main reason for its high growth rate is the wide variety of properties[1,2] which can be achieved for example by incorporating comonomers to give copolymers with decreased melting points down to 120 °C or to give impact modified polypropylenes with rubber contents up to 60% directly synthesized in the reactor[3].

But even without using comonomers, polypropylene homopolymers offer a wide variety of properties in terms of molecular weight, melt flow properties, tacticity, stiffness etc..

One of the most important developments to achieve this flexibility in product properties in an economic way is the commercialization of high mileage supported Ziegler-Natta catalysts by several companies[4,5,24]. This paper will demonstrate the capability of this class of catalysts in the production of polypropylene homopolymers and will give a deeper insight into the interdependence between catalyst performance, process parameters and product properties.

CATALYTIC SYSTEM

The catalytic system used in this study[5] is $TiCl_4$ supported on $MgCl_2$ with a phthalate[25] as an internal donor to improve stereospecifity and productivity. To overcome possible problems with powder morphology in the reactor, that means greater fractions of fine and/or coarse material, these chemicals are impregnated on silica which serves as a morphology support whereas in other systems the catalyst is prepolymerized[6,13] to optimize powder morphology.

---

* To whom correspondence should be adressed

New Advances in Polyolefins, Edited by
T.C. Chung, Plenum Press, New York, 1993

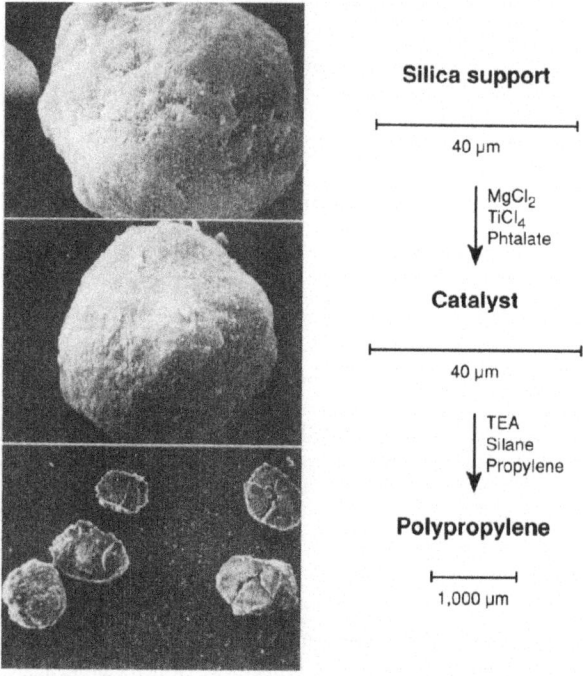

Silica support

|———————————|
40 μm

↓ MgCl₂
  TiCl₄
  Phtalate

Catalyst

|———————————|
40 μm

↓ TEA
  Silane
  Propylene

Polypropylene

|——————|
1,000 μm

Fig. 1. Preparation of supported catalysts on silica and preservation of morphology during polymerization.

Fig. 1 shows the replication of the catalyst morphology throughout the process. The catalyst components are fully absorbed within the silica. The homogeneous distribution of the catalyst components on the support enables an exact replication of the catalyst morphology. Triethylaluminia (TEA) is used as the cocatalyst and a wide variety of dialkyldialkoxysilanes[7,8,26] can be used as external stereomodifiers.

STIRRED BED GAS PHASE REACTORS

The runs described in this paper are from continous plants as shown in fig. 2. The reactors in this plant can be used as stand-alone reactors or as a cascade[9] for example to produce impact modified polypropylene if ethylene is fed to the 2. reactor or to produce homopolymers. Due to the cooling system shown here for the second reactor a wide range of gas compositions can be handled easily.

MOLECULAR WEIGHT VARIATION

One of the main parameters for the application of PP homopolymers is its molecular weight and there exist several applications far above and below the molecular weight range of normal grades. Table 1 gives a rough classification according to applications or processing.

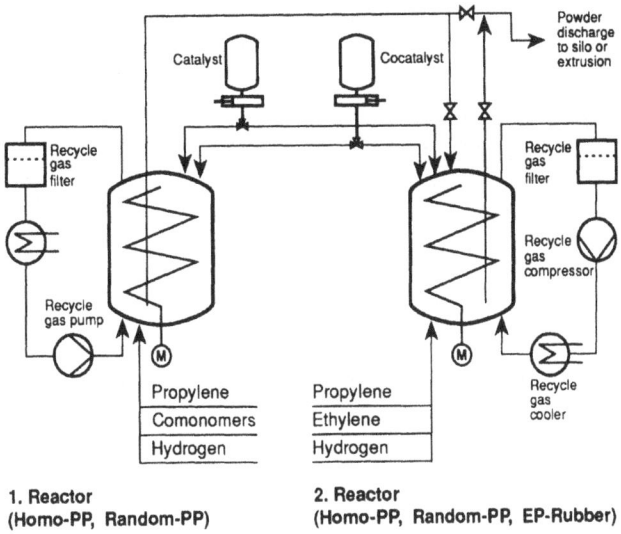

Fig. 2. Typical plant for the stirred bed gas phase process.

Table 1. Applications of PP homopolymers according to M_w.

M_w [g/mol]	application / processing
< 40.000	wax, toner, pigment batch
100.000 - 125.000	melt blown fibres
150.000 - 300.000	injection moulding, fibres
250.000 - 500.000	extrusion, blow moulding, tapes
> 400.000	high melt strength, starting material for CR-PP

Process Parameters and Molecular Weight

There are several variables which may influence the molecular weight, but for a given catalyst system the main parameters are temperature and hydrogen concentration[10,11]. Fig. 3 shows the influence of these process parameters on the molecular weight of polypropylene and on the polydispersity of the resulting molecular weight distribution.

Interestingly, the effect of temperature is relatively small except for low concentrations of hydrogen and the effect levels off for high concentrations. Overall, there is a range of nearly two decades in molecular weight which can be covered with one single catalyst, though very extreme reactor conditions have to be used for the high and low molecular weight products. So hydrogen concentration is between 0 and 40% by volume and temperatures are up to 100 °C.

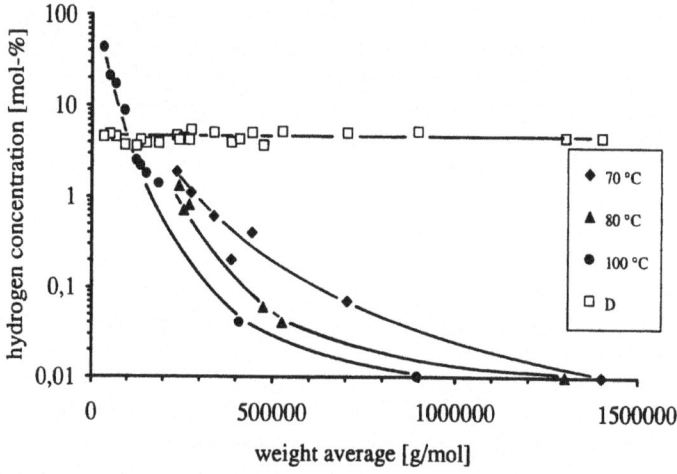

Fig. 3. Influence of hydrogen on molecular weight and polydispersity.

These very extreme conditions offer a unique possibility to look for correlations between reactor conditions, process and catalyst performance and product properties over a wide range.

Due to the extreme hydrogen concentrations, which can easily be achieved in the gas phase process, while maintaining the homogenity of the reaction mixture in contrast to other processes[13], one can examine the dependence of molecular weight on hydrogen concentration over such a wide range. To our knowledge, this has not been considered before in continous reactors.

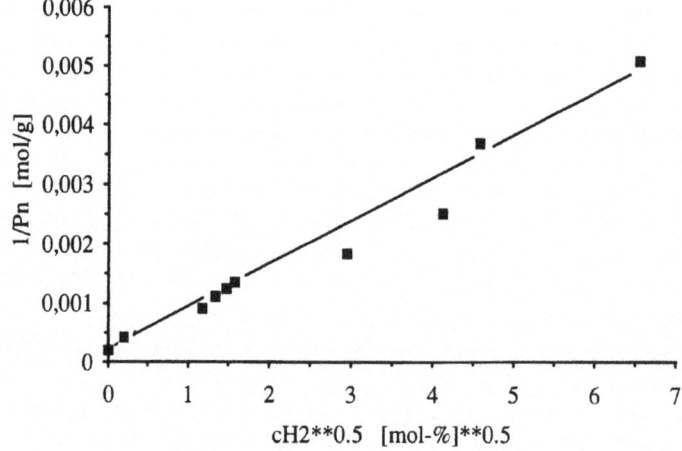

Fig. 4. Reciprocal dependence of P_n on square root of hydrogen concentration.

Over the entire range there is a reciprocal dependence (Fig. 4) of the number average degree of polymerization on the square root of hydrogen concentration,

$$1/P_n = a + b * c_{H2}^{0.5}$$

which clearly supports a mechanism, where hydrogen attack during chain transfer is by adsorbed hydrogen atoms and not by hydrogen molecules[12].

Process Parameters and Performance of Catalyst and Process

The performance of catalyst and process is mainly characterized by two factors: the productivity of the catalyst in the process and the processibility of the product, which is determined by the handling of the reactor powder, which in turn depends on the particle size and particle size distribution.

Fig. 5 shows, that for a given temperature, productivity runs through a maximum, which is surprisingly sharp. It lies somewhere between 150.000 and 400.000 g/mole. Except for molecular weights far below 100.000 g/mol, productivity lies above 10.000 g/g even for very high molecular weights, and is well above 20.000 for molecular weights corresponding to melt flow rates between 2 und 20 g/10min, which are the most common grades.

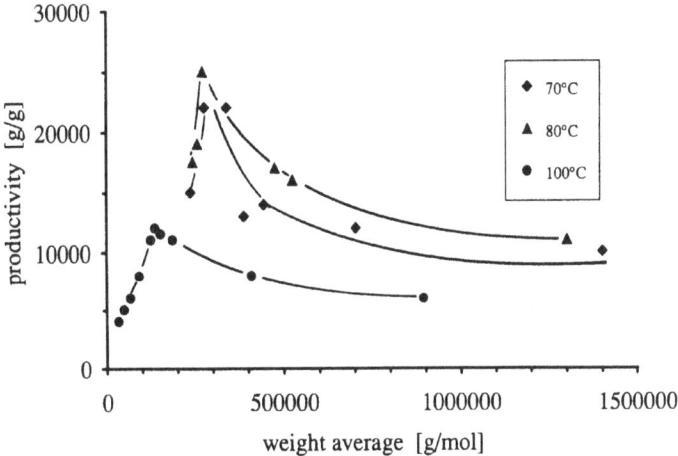

Fig. 5. Dependence of catalyst productivity on molecular weight at different temperatures.

The sharpness of the maxima does not markedly change if one considers the dependence of productivity on hydrogen concentration (fig. 6). In terms of hydrogen concentration they are close together within a very narrow window between 0.5 and 2 % by volume. The productivity drops markedly at somewhat higher or lower concentrations but even for the most extreme case of 40 % H_2 at 100 °C the productivity for the supported catalyst is still on a level, which is at least comparable or even higher than pro-

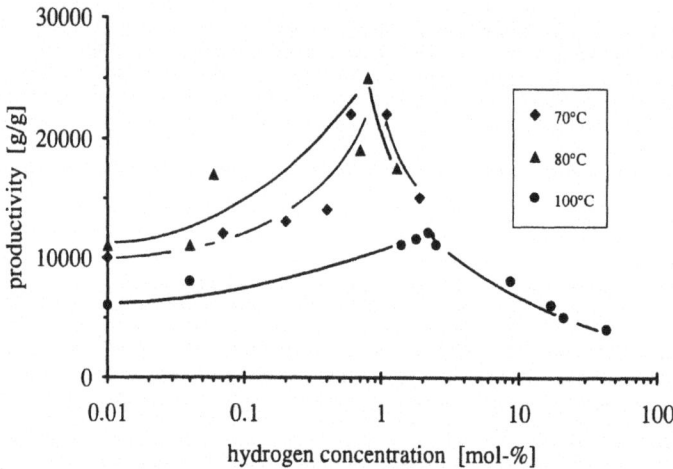

Fig. 6. Dependence of productivity on hydrogen concentration at different temperatures.

ductivity of ball-milled catalysts for products with ususal mel flow rates. The occurence of a maximum in productivity, when varying the hydrogen concentration, can be explained by the kinetic formalism proposed by Keii[11].

The mean particle size, or for better comparison, the 3rd power of the mean particle size (fig. 7) correlates strongly with the productivity as can be expected for catalyst replication during polymerization. Despite this great variation of particle size over the range of molecular weights, the resulting powder does not show difficulties in handling, because the particle size distribution is rather narrow (fig. 8) without fines.

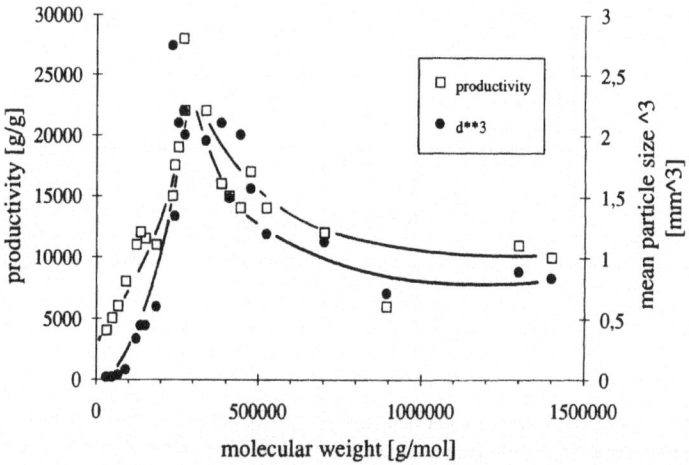

Fig. 7. Comparison of polymer particle size and productivity.

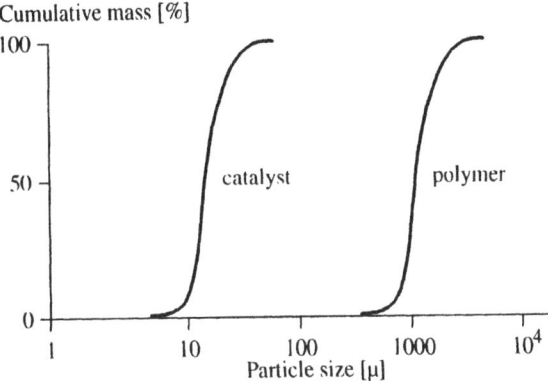

Fig. 8. Comparison of particle size distribution of catalyst and polymer.

Molecular Weight and Product Properties

One important question is, how the other product properties do vary with molecular weight and here we have an excellent opportunity to look at these dependencies, because all products are produced with the same catalyst without any fractionation of the polymer. Only temperature and hydrogen concentrations are changed.

From fig. 3 we can see, that despite the great variations in process parameters, the polydispersion index $D=M_w/M_n$ of the molecular weight distribution remains nearly constant over the whole range of molecular weights and lies between 4 and 5, which is also shown by Keii[12].

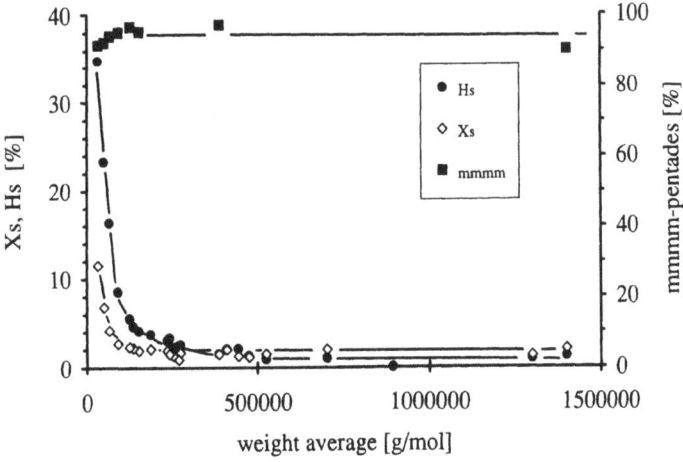

Fig. 9. Dependence of tacticity parameters on molecular weight.

Fig. 9 summarizes the dependence of the properties characterizing the tacticity of the polymer on the molecular weight. The heptane soluble fraction (Hs) remains constant above 500.000 g/mole, but below this value it starts to increase in an exponential way up to 34 % for the wax like polymers. The xylene soluble fraction remains constant down to 100.000 g/mole where it starts to increase sharply but to lower values than the Hs. Contrary to this behaviour the concentration of mmmm-pentades is rather constant over the whole range.

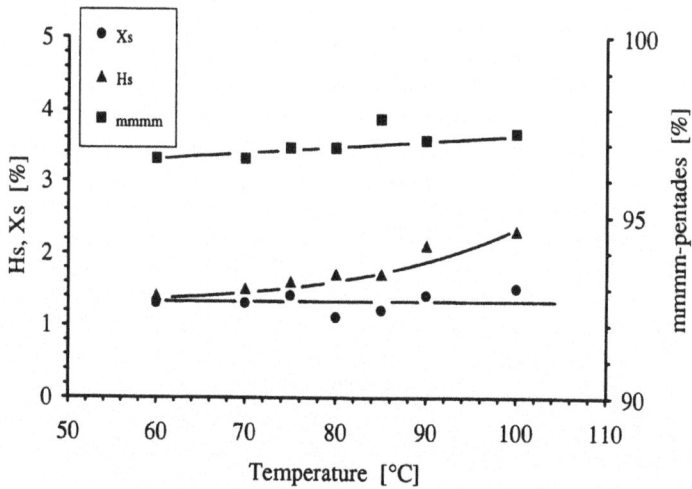

Fig. 10. Dependence of tacticity parameters on temperature.

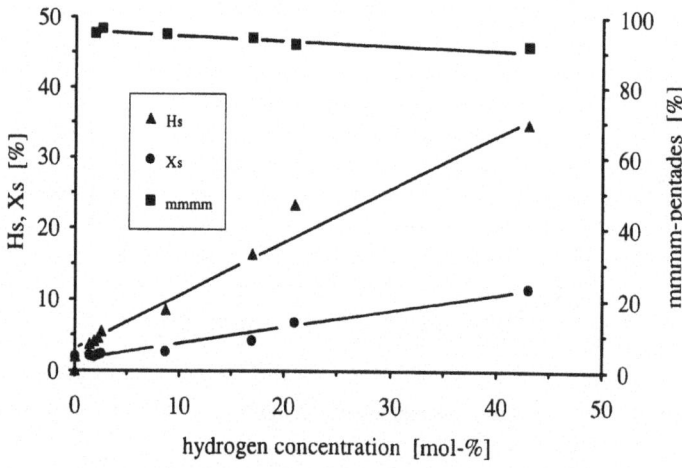

Fig. 11. Dependence of tacticity parameters on hydrogen concentration.

To get some better understanding, figs. 10 and 11 show the dependence of these parameters on temperature and hydrogen concentration separately.

The detailed information for temperature dependence is from batch experiments. The Xs is constant, whereas the Hs and the mmmm-pentades increase slightly (less than 1%) with temperature. So temperature has only a very small effect on tacticity.

The effect of hydrogen[6] seems to be somewhat greater; by increasing the hydrogen concentration by a factor of 40 both soluble fractions increase markedly whereas the mmmm-pentades are lowered from 96 to 89 %, but this cannot be the reason for the dramatic effect on soluble fractions because high molecular weight products (fig. 9) with the same mmmm-fraction show much lower solubles.

So despite the great variation in process parameters, there are generally only minor changes in tacticity of the polymer. This means, the catalyst is very stable against changes in reactor conditions.

The high tacticity of the product irrespectively of its molecular weight is also reflected by the high cristallinity of the polymer of more than 70 % (fig. 12), which increases with decreasing molecular weight due to the higher mobility of shorter chains. The melting point generally lies at 165-167 °C and decreases when the molecular weight drops below 100.000g/mol[14].

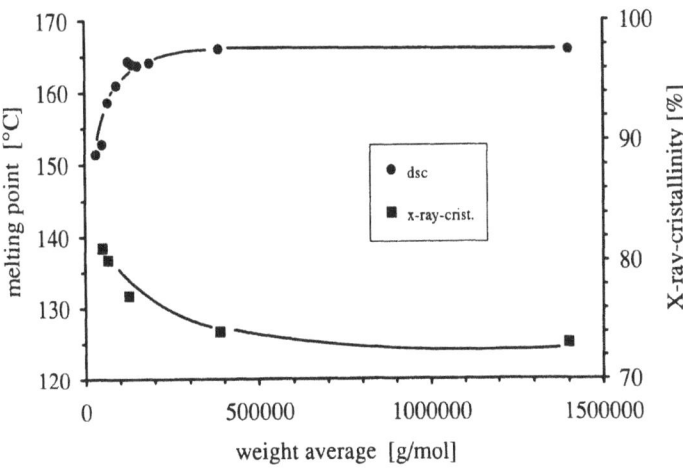

Fig. 12. Dependence of cristalline properties on molecular weight.

The lamellar structure of the polymers changes with molecular weight, as can be seen from fig. 13. With decreasing molecular weight the length of the lamellars increase as an effect of the higher mobility. For wax-like polymers with M_w=30.000 g/mol a pronounced cross-hatched structure can be observed[15]. The angle of about 81° between the crossing lamellars prevents a complete occupation of the lattice within the polymer, so there are voids within the polymer structure. These voids and the lack of molecules long enough to accomplish bridging between the spherulites are responsible for the high brittleness of these low molecular products.

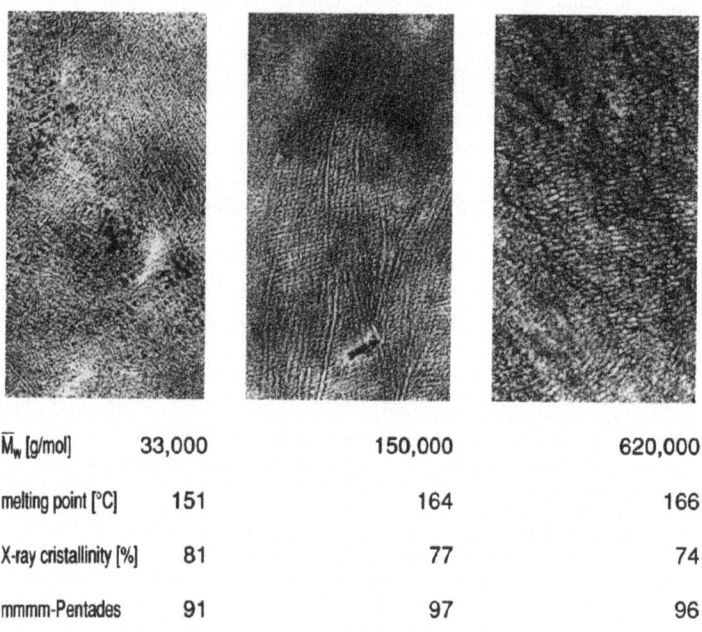

$\overline{M}_w$ [g/mol]	33,000	150,000	620,000
melting point [°C]	151	164	166
X-ray cristallinity [%]	81	77	74
mmmm-Pentades	91	97	96

Fig. 13. Lamellar structure of polypropylenes with different molecular weights.

Mechanical Properties and Molecular Weight

The mechanical properties of homopolypropylene, which are most interesting for the customer, are stiffness and impact strength. The dependence of these parameters on molecular weight can be seen in fig. 14. For products with $M_w \ll 100.000$ g/mol no reliable data are available because these polymers are too brittle to get reproducible results. For all other products, the shear modulus decreases sharply with increasing molecular weight from 900 N/mm² down to 600 N/mm² for molecular weights above 1.000.000 g/mol. Correspodingly impact strength, both notched and unnotched, decreases with decreasing molecular weight.

One reason for this behaviour is the dependence of cristallinity on molecular weight. Decreasing the molecular weight results in higher cristallinity and also leads to higher stiffness and lower impact strength because of decreasing amorphous fraction.

TACTICITY VARIATION

Besides the molecular weight, the tacticity of the polymer is an important property. As we saw before, this property is insensitive against variations in process parameters like temperature and hydrogen concentration, but it can be easily varied by the external stereomodifier, while leaving the catalyst and the cocatalyst unchanged. There are se-

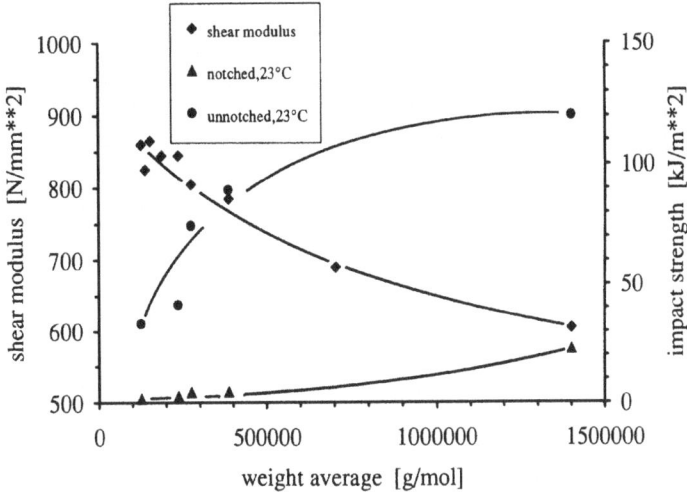

Fig. 14. Dependence of mechanical properties on molecular weight

veral publications [7,8,16] dealing with the influence of the structure of the silane on catalyst stereospecifity and the polymer tacticity, too. For practical purposes however, a more elegant way for influencing the stereospecifity of a catalytic system is to change the silane concentration and to adjust the tacticity to the desired value simply by regulating one pump. This possibility of varying the catalyst specifity independently of nearly all other parameters is one of the main advantages of the supported catalysts.

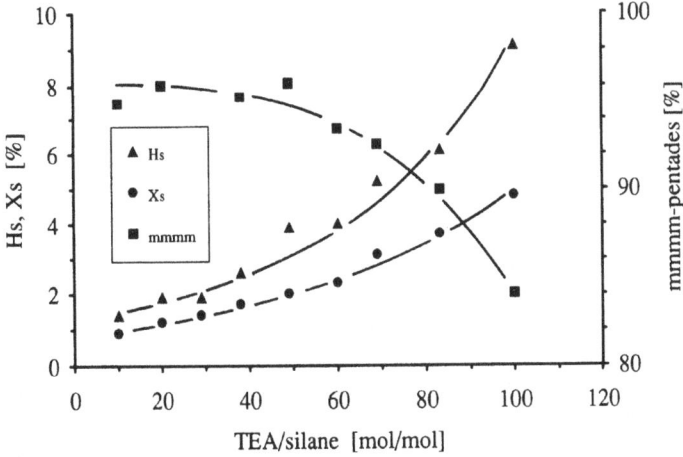

Fig. 15. Variation of tacticity parameters with the concentration of the external donor.

Silane Concentration and Tacticity Parameter

Fig. 15 shows the dependence of properties characterizing the tacticity of the polymer on the silane concentration, which is given as the ratio of cocatalyst to silane. The molecular weight of all products is adjusted to give an mfr of 8 g/10min.

With constant molecular weight both soluble fractions, Xs as well as Hs, increase correspondingly. Only for a constant M_w it is possible to characterize the catalyst stereospecifity by both parameters; for changing molecular weights however, as we saw before, the Xs is the better suited parameter.

Silane Concentration and Application Properties

With changing tacticity also the macroscopic properties of the polymer change (fig. 16). There is only a minor effect on the melting point; it drops slightly from 166 to 164°C. However, the mechanical properties change to a greater extent. So for example, there is a marked increase in stiffness, when increasing the silane concentration by 33%. The reason for this is a decrease of the amorphous fraction.

Silane Concentration and Process Parameters

All these variations can be done simply by changing the silane concentration, using the same catalyst and without altering cocatalyst concentration, pressure, temperature etc.. The only parameter which has to be adjusted slightly is the hydrogen concentration in the reactor to keep thr mfr constant.

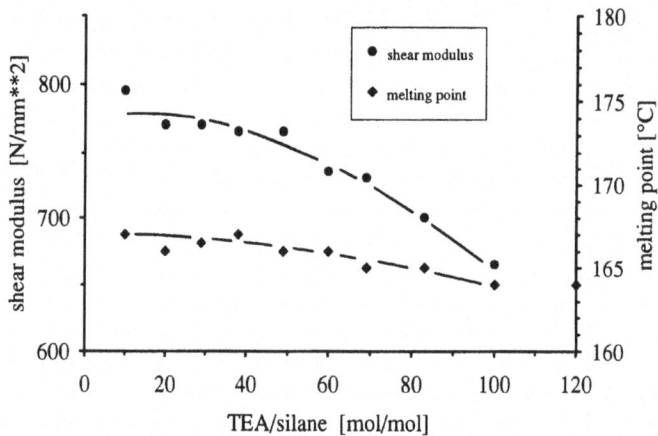

Fig. 16. Influence of silane concentration on macroscopic properties.

Usually the more stereospecific catalyst gives a product with a higher molecular weight[7] and therefore a higher hydrogen concentration in the reactor has to be maintained to keep the mfr constant for the runs with a higher silane concentration (fig. 17).

The decrease in hydrogen concentration for lower silane concentrations is the reason for the slight drop in productivity when lowering polymer tacticity. From fig. 4 we saw that the maximum in productivity is somewhere around 1 % hydrogen and decreases for lower concentrations. This is exactly that, what we have to do, if we change the ratio TEA/silane from 10 to 100.

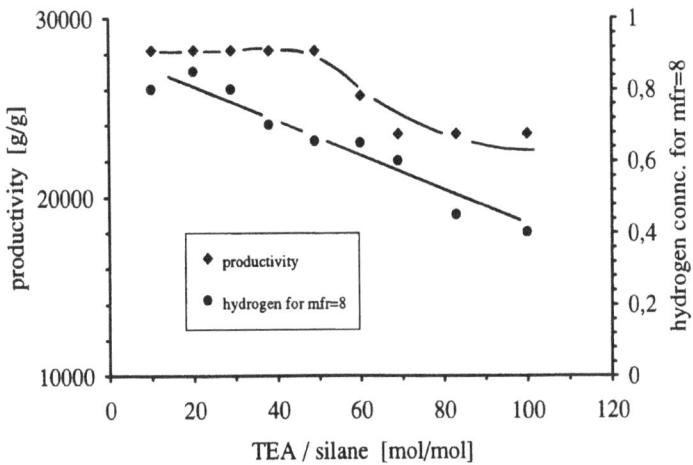

Fig. 17. Influence of silane concentration on process and catalyst performance.

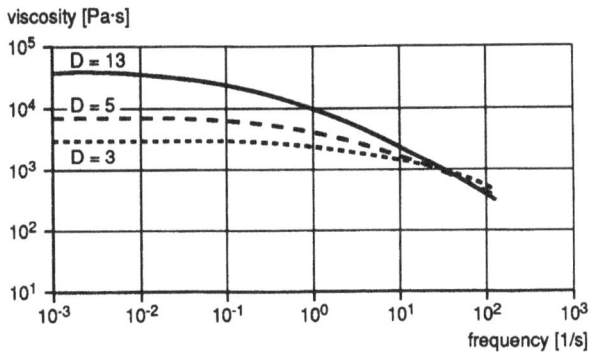

Fig. 18. Variation of melt flow properties with mwd for PP-grades with mfr=5.

VARIATION OF MOLECULAR WEIGHT DISTRIBUTION

The third important property of homopolypropylene is the shape and width of its molecular weight distribution (mwd). This is the only property which cannot be influenced by catalyst or process parameters to a considerable extent, and so the process itself has to be changed. For narrowing the mwd, peroxide degradation[18,19] is a well accepted method, and to broaden the molecular weight distribution the use of supported catalyts in a reactor cascade[17] is the most promising method. In both cases, the polymer properties arising from the catalyst remain unchanged.

The main reason for producing polypropylenes with different molecular weight distributions is the dependence of melt flow properties[20,21] and processing properties[22,23] on the width and shape of the mwd (fig. 18).

CONCLUSION

Though Ziegler - Natta Catalysis now is almost 40 years old, the class of supported catalysts in combination with appropiate process parameters make accessible a wide range of homopolymers in terms of:

> molecular weight
> tacticity
> mwd

which can be chosen independently to meet the demand for commodities and specialities.

REFERENCES

1) P. Galli, J.C. Haylock, Continuing initiator system developments provide a new horizon for polyolefin quality and properties, *Prog. Polym. Sci.*,16:443 (1991)
2) P. Galli, F. Milani, T. Simonazzi, New trends in the field of propylene based polymers, *Polymer.*,17:37 (1985)
3) E. Seiler, B. Göller, Polypropylen, *Kunststoffe*, 80:1085 (1990)
4) EP 45977
5) EP 288845, EP 306867
6) T. Yano, T. Inoue, S. Ikai, Y. Kai, M. Tamura, M. Shimizu,.Two-step polymerization of propene by magnesium supported catalysts ,*Makromol.Chem., Rapid Commun.* 7:491(1985)
7) T. Miyatake, K. Mizunuma, M. Kakugo, Microtacticity distribution of polypropylenes with MgCl$_2$ supported Ti catalyst systems, *in*: "Catalytic Olefin Polymerization" T. Keii, K. Soga, ed.,Elsevier, Oxford (1990)
8) J.V. Seppälä, M. Härkönen, Effect of the structure of external alkoxysilane donors on the polymerization of propene with high activity Ziegler-Natta catalysts, *Makromol. Chem.*, 190:2535 (1989)
9) DE 3244312, DE 3324793
10) P.C. Barbé, G. Cecchin, L. Noristi, The catalytic system Ti-complex/MgCl$_2$, *Adv.Polym.Sci., 81:1 (1987)*
11) T. Keii, Fundamental aspects of hydrogen on Ziegler-Natta polymerization, *in*: "Transition Metal Catalyzed Polymerizations", R.P. Quirk, ed., Cambridge University Press, Cambridge (1988)
12) T. Keii, Y. Doi, E. Suzuki, M. Tamura, M. Murata, K. Soga, Propene polymerization with a MgCl$_2$ supported Ziegler catalyst, *Makromol. Chem.*,185:1537 (1984)

13) T. Simonazzi, G. Cecchin, S. Mazzullo, An outlook on progress in polypropylene based polymer technology, *Prog. Polym. Sci.*, 16: 303 (1991)

14) G. Natta, I. Pasquon, A. Zambelli, G. Gatti, Dependence of the melting point of isotactic polypropylenes on their molecular weight and degree of stereospecifity of different catalytic systems, *Makromol. Chem.*,70:191 (1963)

15) D. R. Norton, A. Keller, The spherulitic and lamellar morphology of melt-crystallized isotactic polypropylene, *Polymer*, 26:704 (1985)

16) M. Härkönen, J. V. Seppälä, External silane donors in Ziegler-Natta catalysis, *Makromol. Chem.*, 192:2857 (1991)

17) K. D. Hungenberg, Strategies for PP-homo- and copolymers with broadened mwd in stirred bed gas phase reactors, *in*: "4th Int. Workshop on Polymer Reaction Engineering", K.-H. Reichardt, H.-U. Moritz ed., VCH, Berlin (1992)

18) J. Curry, S. Jackson, B. Stoehrer, A. van der Veen, Free radical degradation of polypropylene, *Chem. Eng. Prog.*, 84:43 (1988)

19) H. G. Fritz, B. Stöhrer, Polymer compounding process for controlled peroxide degradation of PP, *Intern. Polym. Processing*, 1: 31 (1986)

20) H.M. Laun, Rheologie von Kunststoffschmelzen mit unterschiedlichem molekularem Aufbau. *Kautschuk+Gummi*, 40:554 (1987)

21) W. Minoshima, J. L. White, J. E. Spruiell, Experimental investigation of the influence of molecular weight distribution on the rheological properties of polypropylene melts, *Polym. Eng. Sci.*, 20:1166 (1984)

22) L. Bourland, Bimodal mwd, polypropylene alloys of lmwpp polymers/conventional PP resins, *ANTEC* 1223 (1988)

23) A. Wolfsberger, Calendering of PP - product requirements, processing and applications, presented at the "Speciality Plastics Conference 1990", Zürich, 1990

24) K.-Y Choi, H.W. Ray, Recent developments in transition metal catalyzed polymerization - A survey, *J. Macromol. Sci.-Rev. Macromol. Chem. Phys.*, C25:57 (1985)

25) K. Soga, T. Shiono, Effect of diesters and organosilicon compounds on the stability and stereospecifity of Ziegler-Natta catalysts, *in* "Transition Metal Catalyzed Polymerizations: Ziegler-Natta and Metathesis Polymerizations", R. P. Quirk, ed., University Press, Cambridge (1988)

26) A. Proto, L. Oliva, C. Pelecchia, A.J. Sivak, L.A. Cullo, Isotactic-specific polymerization of propene with supported catalysts in the presence of different modifiers, *Macromolecules*, 23:2904 (1990)

PALLADIUM(II)-CATALYZED ALTERNATING COPOLYMERIZATION AND TERPOLYMERIZATION OF CARBON MONOXIDE WITH α-OLEFINS

Zhaozhong Jiang, Garth M. Dahlen, and Ayusman Sen*

Department of Chemistry, The Pennsylvania State University
University Park, PA 16802

INTRODUCTION

The copolymers of olefins with carbon monoxide are of great interest from at least three standpoints.[1] First, as a monomer, carbon monoxide is particularly plentiful and inexpensive. Second, the presence of the carbonyl chromophore in the backbone makes them photodegradable.[2] A third reason for the interest in olefin-carbon monoxide copolymers is that, because of the ease with which the carbonyl group can be chemically modified, the polyketones serve as excellent starting materials for other classes of functionalized polymers. In fact, about two dozen polymers incorporating a variety of functional groups have been previously synthesized[1] from the *random* ethylene-carbon monoxide copolymer (C_2H_4:CO > 1) made through radical-initiated polymerization. Since carbon monoxide does not homopolymerize, the *alternating* olefin-carbon monoxide copolymers (olefin:CO = 1) has the highest possible concentration of the reactive carbonyl groups. Additionally, the 1,4-arrangement of the carbonyl groups in the alternating olefin-carbon monoxide copolymers provides additional functionalization pathways[3] (Fig. 1). For example, starting with the alternating ethylene-carbon monoxide copolymer, we have synthesized novel polymers with interesting electrical properties.[3b]

Three different synthetic methods have been developed for the copolymerization of olefins with carbon monoxide.[1] The copolymerization may be initiated by free radicals, or induced by γ-rays. A number of transition metal complexes are also effective catalysts for the copolymerization reaction. The metal-catalyzed copolymerization procedure has several noteworthy advantages over the other two. First, the copolymerization occurs at significantly lower pressures and at or below ambient temperature. Second, the resultant polymers have a strictly alternating structure and high molecular weights.

The metal-catalyzed alternating copolymerization of carbon monoxide with ethylene and the properties of the resultant ethylene-carbon monoxide copolymer (E-CO copolymer) have been described in detail by us[4] and others.[5] Much less is, however, known regarding the alternating copolymers of α-olefins with carbon monoxide. Most of the work in the latter area is in the patent literature[6] although a few papers have appeared very recently.[7] In this chapter, we intend to systematically describe several catalytic systems for the synthesis of alternating copolymers and terpolymers of carbon monoxide with α-olefins that are based on cationic, tertiary phosphine complexes of palladium(II). Additionally, we will discuss the initiation, chain propagation, and termination steps that are involved in the copolymerization. Finally, we show how the above mechanistic information can be used to rationally design catalysts for the alternating copolymerization of olefins with carbon monoxide in the absence of a solvent.

SYNTHESIS OF ALTERNATING COPOLYMERS OF CARBON MONOXIDE AND α-OLEFINS

The compound, $[Pd(Ph_2P(CH_2)_3PPh_2)(MeCN)_2](BF_4)_2$, **1**, was prepared in situ by codissolving a 1:1 (molar ratio) mixture of $[Pd(MeCN)_4](BF_4)_2$[8] and $Ph_2P(CH_2)_3PPh_2$ (dppp) in an appropriate solvent. In a number of solvents, **1** was found to be a highly active catalyst for the alternating copolymerization of CO with olefins.[7a] For example, in a nitromethane/methanol (2:1 v/v) mixture, using 500 psi each of C_2H_4 and CO, the E-CO copolymer was formed at the initial rate of 3.6 Kg copolymer/g.Pd-h when the reaction was carried out at 66°C, (productivity over 19 h at 70°C: 28 Kg/g.Pd). The addition of water (49 molar equivalents per Pd) and 2,6-di-tert-butylpyridine (8 molar equivalents per Pd) did not affect the reaction rate. No decomposition of the catalyst was observed at the end of the reaction and the catalyst could be used repeatedly. The E-CO copolymer was a white crystalline solid insoluble in common organic solvents, but soluble in strong acids, such as 1,1,1,3,3,3-hexafluoroisopropanol and trifluoroacetic acid [^{1}H-NMR (CF_3COOD) (ppm): 3.02 (s); ^{13}C{^{1}H}NMR (CF_3COOD) (ppm): 217.02, 38.03].

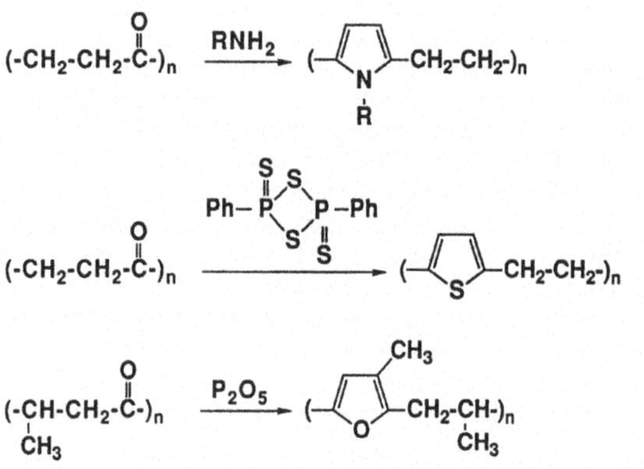

Figure. 1. Representative Examples of Functionalized Polymers Derived from the Alternating Olefin-Carbon Monoxide Copolymers by the Reaction of the 1,4-Carbonyl Groups.

The choice of solvent had a dramatic effect on the rate of reaction. A polar solvent system with an alcohol as one component, such as nitromethane/methanol mixture, appeared to be optimal. In the absence of alcohol, a long induction period followed by a slower rate was observed. For example, in THF/nitromethane mixture, a 2 h induction period was observed.

Compound **1** was also found to catalyze the copolymerization of carbon monoxide with α-olefins and α,ω-dienes, such as propylene, 1-heptene, and 1,7-octadiene. Starting with approx. 30 g of C_3H_6 and 600 psi of CO, the P-CO copolymer was formed at the average rate of 48 g copolymer/g.Pd-h at 50°C in a 2:1 (v/v) mixture of methanol/nitromethane. The copolymerization of CO with 1-heptene or 1,7-octadiene was performed following a similar procedure employed for propylene. 1-Heptene showed a significantly lower reactivity than propylene in the copolymerization reaction due to the following reasons. First, because of its greater steric bulk, 1-heptene would be expected to insert more slowly into the intermediate metal-acyl species in the chain-growth sequence. Second, 1-heptene was found to be isomerized in the absence of CO to form internal olefins[9] that were no longer able to copolymerize with CO under the same reaction conditions as demonstrated by control experiments. The alternating copolymers of carbon monoxide with propylene and 1-heptene had molecular weights of $3.1\text{-}2.4 \times 10^4$ and 1×10^3, respectively, versus standard polystyrene. Both copolymers were soluble in common organic solvents such as CH_2Cl_2, $CHCl_3$, and THF.

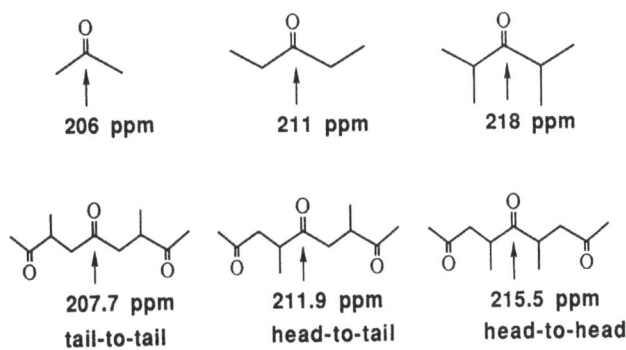

Figure. 2. A Comparison of the ^{13}C-NMR Resonances for the Carbonyl Groups in Model Compounds and in the Repeating Units of the Alternating Propylene-Carbon Monoxide Copolymer.

The P-CO copolymer was an atactic alternating copolymer with a mixture of head-to-tail and head-to-head repeating units. The characterization of the copolymer was accomplished by IR, ^1H- and ^{13}C-NMR spectroscopy and supported by elemental analysis. A strong IR absorption (neat) at 1710 cm^{-1} was ascribed to the carbonyl group. The ^1H-NMR ($CDCl_3$) spectrum of the copolymer exhibited three broad resonances at 2.89-3.16 ppm (2H), 2.33-2.50 ppm (1H), and 1.01-1.10 ppm (3H) corresponding to the CH_2, CH, and CH_3 groups, respectively. The $^{13}C\{^1H\}$NMR ($CDCl_3$) spectrum displayed multiplets at 215.46, 211.91, 207.70, 44.62, 40.29, and 16.34 ppm. The first three resonances are due to the carbonyl group and the latter were attributed to the CH_2, CH, and CH_3 groups, respectively. The multiplicity of the resonances was caused by the lack of tacticity and the presence of a mixture of head-to-tail and head-to-head repeating units in the the polymer backbone. We assign the absorptions at 215.46, 211.91, 207.70 ppm to carbonyl groups flanked by head-to-head, head-to-tail, and tail-to-tail propylene units based on the ^{13}C-NMR resonances of the carbonyl groups of model compounds as shown in Fig. 2. The intensity ratios of the carbonyl resonances of the P-CO copolymer indicated that the copolymer has 60% head-to-tail, 20% head-to-head, and 20% tail-to-tail arrangements.

As expected, cross-linked polymers were obtained in the reaction of α,ω-dienes with CO. For example, the 1,7-octadiene-CO copolymer was a white solid that was totally insoluble in organic solvents but swelled in solvents such as THF. $^{13}C\{^1H\}$NMR spectra revealed the presence of carbonyl groups, unreacted terminal vinyl groups, internal double bonds, and methoxycarbonyl groups in addition to the alkyl groups. The internal double bonds were formed through the isomerization of the pendant vinyl groups, while the methoxycarbonyl groups were formed due to chain initiation by Pd-methoxide (vide infra).

Owing to the versatility of the present catalyst system, it was possible to prepare different kinds of olefin-carbon monoxide copolymers in one pot or to incorporate more than one olefin into the copolymer chains. For example, one pot synthesis of the E-CO and the P-CO copolymers was accomplished by the following procedure. A solution containing 1.13×10^{-2} mmol of the catalyst in 1.2 ml of 2:1 (v/v) nitromethane/methanol was exposed to a mixture of C_2H_4 (500 psi) and CO (500 psi) at room temperature for 20 h. At the end of this period, the gases were expelled and the apparatus (with its contents untouched) was recharged with a mixture of C_3H_6 (30 g) and CO (600 psi) at 50°C. After 3 d, 12.4 g solid was obtained, which was separated into pure E-CO copolymer (6 g, insoluble) and pure P-CO copolymer (6.4 g, soluble) by extraction with CH_2Cl_2 and identified by NMR spectroscopy. On the other hand, a terpolymer containing *both* alternating E-CO and alternating P-CO units[10] was formed when the catalyst solution in nitromethane/THF was exposed to a mixture of C_2H_4, C_3H_6 and CO. The ^1H-NMR (CF$_3$COOD) spectrum of the polymer showed absorptions at 2.93 (s), 3.2-2.6 (br), and 1.19 (d, J = 6.5 Hz) ppm due to the -CH$_2$CH$_2$- group of the E-CO units, and -CH$_2$CH- and -CH$_3$ groups of the P-CO units, respectively. The terpolymer had 14% P-CO units and was insoluble in common organic solvents, but appeared to swell in CH_2Cl_2. Interestingly, only E-CO copolymer was formed when the reaction was performed in nitromethane/methanol rather than in nitromethane/THF. Thus, it was possible to tailor catalyst selectivity by the appropriate choice of the solvent.

Since the copolymerization of the α,ω-dienes with CO led to the formation of cross-linked materials, they were good candidates as cross-linking reagents for the copolymerization of CO with α-olefins. Thus, when 1,7-octadiene was added during the copolymerization of C_3H_6 and CO, a cross-linked P-CO copolymer was obtained. This polymer had low solubility in organic solvents and formed a gel in CHCl$_3$.

Regioregular Control in the Alternating α-Olefin-CO Copolymerization Reaction

It was possible to control the regioregularity in the structure of the α-olefin-CO copolymers by the proper choice of phosphine ligands attached to the Pd(II) ion. For example, with Ph$_2$P(CH$_2$)$_3$PPh$_2$ (dppp) as the ligand, the P-CO copolymer formed always had a structure consisting of head-to-head, head-to-tail, and tail-to-tail arrangements in the polymer backbone. However, the head-to-tail structure predominated when more basic phosphines, such as 1,2-bis(dicyclohexylphosphino)ethane and trimethylphosphine, were used as ligands. This was shown by the ^{13}C-NMR(CDCl$_3$) spectrum of the resultant P-CO copolymer which exhibited only the head-to-tail carbonyl resonance at 211.96 ppm. It is clear from the structure of the phosphines employed that the regioregularity of the resultant copolymer was a function of the basicity rather than the steric size of the ligand.

The compound, [Pd(PMe$_3$)$_2$(MeCN)$_2$](BF$_4$)$_2$, was also active for the copolymerization of styrene and carbon monoxide to form low molecular weight atactic alternating styrene-carbon monoxide (S-CO) copolymer. The ^1H-NMR (CDCl$_3$) spectrum of the S-CO copolymer exhibited broad resonances at 7.2 (5H) and at 4.0 (1H), 3.1 (1H), and 2.6 (1H) ppm due to the phenyl and the -CH(Ph)-CH$_2$- groups, respectively. The $^{13}C\{^1H\}$NMR (CDCl$_3$) spectrum showed resonances as multiplets at 206.27 (carbonyl), 137.23, 126.20-128.81 (phenyl), 52.43 (CH), and 44.33 (CH$_2$) ppm. The single carbonyl absorption at 206.27 ppm indicated a prevailing head-to-tail structure for the copolymer.

MECHANISTIC ASPECTS OF THE ALTERNATING COPOLYMERIZATION OF CARBON MONOXIDE WITH OLEFINS

By using $[Pd(PPh_3)_2(MeCN)_2](BF_4)_2$, and related compounds as catalysts, we had earlier demonstrated that, for the E-CO copolymer, the chain-growth occurred by alternate insertions of CO and olefin into an initial Pd-alkyl bond (Scheme 1).[4b] Two possible factors favor CO insertion into a Pd-alkyl bond over the corresponding olefin insertion. First, because of its greater binding ability, the local concentration of CO was expected to be significantly higher than that of the olefin. In addition, CO appears to have a greater inherent tendency to insert into transition metal-alkyl bonds than do olefins. Note that in the absence of added CO, the same catalyst system was effective for the dimerization of olefins under identical conditions.[4] This clearly indicated that there was no thermodynamic barrier to olefin insertion into the Pd-alkyl bonds in this system; rather, the olefin insertion was too slow to compete with the corresponding CO insertion. The formation of olefin dimers rather than polymers in the absence of CO indicated the presence of a facile β-hydrogen abstraction pathway. This pathway was also found to be one of the chain termination/transfer steps in the alternating olefin-carbon monoxide copolymerization (vide infra). In contrast to insertions involving Pd-alkyl bonds, insertions into Pd-acyl bonds were dictated by thermodynamics. As shown in Scheme 2,[11b] our work on α-ketoacyl complexes of Pd(II) clearly indicated that CO insertion into a Pd-acyl bond was "uphill" from a thermodynamic standpoint.[11] Thus, only olefin insertion into Pd-acyl bonds was observed.

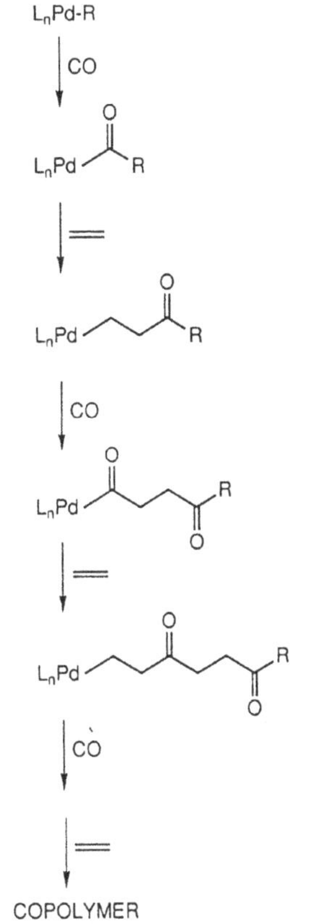

SCHEME 1

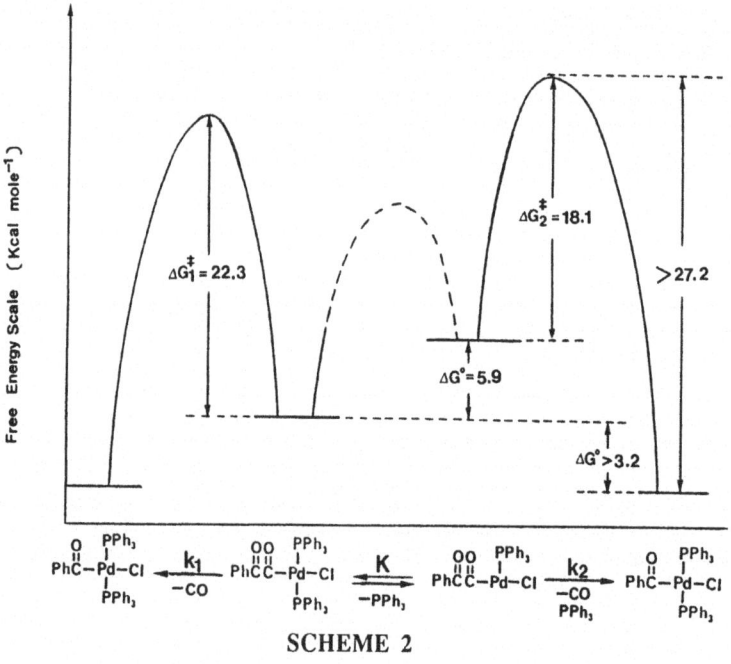

SCHEME 2

While the chain-growth mechanism was established for the alternating E-CO copolymers, there was no compelling reason to postulate a different mechanism for the copolymerization of CO with α-olefins. However, it was important to establish the steps by which the polymer chains were initiated and terminated since these may vary with the particular catalyst and solvent system employed. In order to examine these steps, the end-groups of the P-CO copolymer were determined in the following way. A low molecular weight P-CO copolymer sample (average MW = 2400) that was synthesized using a nitromethane/methanol solvent mixture was heated at 140°C under 2.5×10^{-5} mmHg and the oligomers that distilled over were trapped in a liquid nitrogen trap. A CHCl$_3$ solution of the oligomers was analyzed by GC-MS and identified as shown in Fig. 3. The structures of the oligomers were further confirmed by ^{1}H-NMR spectroscopy. The majority of the oligomers contained alkyl-alkyl or alkyl-propenyl end-groups. These oligomers were clearly formed as a result of reactions initiated by a Pd-hydride and terminated by either proton cleavage of or β-hydrogen abstraction from the final Pd-alkyl species. A relatively smaller number of the oligomers were found with methoxycarbonyl-alkyl or methoxycarbonyl-propenyl end-groups. The former could have been initiated by a Pd-hydride and terminated by methanolysis of a Pd-acyl species (as was observed previously[4b]) *or* initiated by a Pd-methoxide and terminated by proton cleavage of a Pd-alkyl species. However, the latter oligomer *must* have formed through initiation by a Pd-methoxide and termination by β-hydrogen abstraction from a Pd-alkyl species. Likewise, the oligomers with methyl ether end-group *must* also be initiated by a Pd-methoxide. In short, both a Pd-hydride *and* a Pd-methoxide must act as initiators. The initial palladium-hydride could be generated through β-hydrogen abstraction from a Pd-alkoxide[12] generated in situ. In support of this assumption, we observed the formation of acetone from isopropanol at ambient temperature in the presence of **1** in CD$_3$NO$_2$. Additionally, the copolymerization yield was significantly lower in tertbutanol/nitromethane mixture when compared to methanol/nitromethane or isopropanol/nitromethane solvent mixtures presumably because of the absence of β-hydrogens in Pd-tertbutoxide. The initiation of the copolymerization by a Pd-alkoxide is also feasible since the insertion of CO and olefins into Pd-OR bonds is well documented.[12] Depending on whether CO or C$_3$H$_6$ inserted first, the copolymer would either have methoxycarbonyl or methyl ether terminal groups; both were observed. Finally, by switching from methanol/nitromethane to isopropanol/nitromethane mixture, isopropoxide end-groups were formed as detected by ^{1}H-NMR spectroscopy of the isolated copolymer. The above conclusions regarding the initiation and termination steps in the formation of the P-CO copolymer also appear to be applicable to the E-CO copolymer. The E-CO copolymer formed using **1** in nitromethane/methanol showed three small ^{1}H-NMR (CF$_3$COOD) resonances at 5.83, 3.75, and 1.22 ppm due to the terminal vinyl, methoxycarbonyl, and methyl (of ethyl) groups. Scheme 3 outlines some of the principal steps involved in the copolymerization reaction in the presence of methanol. Drent[5b] and Consiglio[7e] have reported similar results for related systems in recent publications.

H-(-CH$_2$-CH-CO-)$_n$-C$_3$H$_7$
  |
  CH$_3$

(61.4%)

H-(-CH$_2$-CH-CO-)$_n$-C$_3$H$_5$
  |
  CH$_3$

(26.9%)

CH$_3$O-CO-(-CH$_2$-CH-CO-)$_n$-C$_3$H$_7$
    |
    CH$_3$

(6.0%)

CH$_3$O-CO-(-CH$_2$-CH-CO-)$_n$-C$_3$H$_5$
    |
    CH$_3$

(3.4%)

CH$_3$O-(-CH$_2$-CH-CO-)$_n$-C$_3$H$_7$
    |
    CH$_3$

(2.3%)

Figure. 3. The Low Molecular Weight Alternating Propylene-Carbon Monoxide Oligomers Formed in Nitromethane/Methanol Mixture.

The initiation of copolymerization by Pd-hydride and Pd-alkoxide formed in situ in alcohol-containing solvents is, perhaps, not too surprising. A more interesting question involves the nature of the initiator in aprotic media. As described earlier, in the absence of alcohol, such as in nitromethane/THF, a long induction period followed by a slower copolymerization rate was observed. The induction period for the copolymerization of C$_2$H$_4$ with CO in nitromethane/THF mixture disappeared upon the addition of 200 psi of H$_2$. Given the high electrophilicity of the Pd(II) ion,[13] the formation of a Pd-hydride by heterolytic cleavage of H$_2$ is not unexpected. That a Pd-hydride was the initiator in the presence of H$_2$ was also indicated by the following observation. The ^{2}H-NMR spectrum (in CF$_3$COOH) of the E-CO copolymer formed in nitromethane/THF in the presence of D$_2$ showed substantial deuteration of the terminal ethyl group (CH$_2$, 2.70 ppm; CH$_3$, 1.18 ppm).The deuteration of the methylene unit of the ethyl terminus was consistent with a rapid reversible insertion of C$_2$H$_4$ into the initial Pd-D bond. In the *absence* of added H$_2$, the Pd-hydride species may form from trace water via the water gas shift reaction.[14] Metal-hydrides are known to be intermediates in this reaction. In support of the above hypothesis, we observed deuteration of the terminal ethyl group of the E-CO copolymer when the polymerization was carried out in nitromethane/THF in the presence of added D$_2$O. One termination step in an aprotic solvent appears to be β-hydrogen abstraction from the final

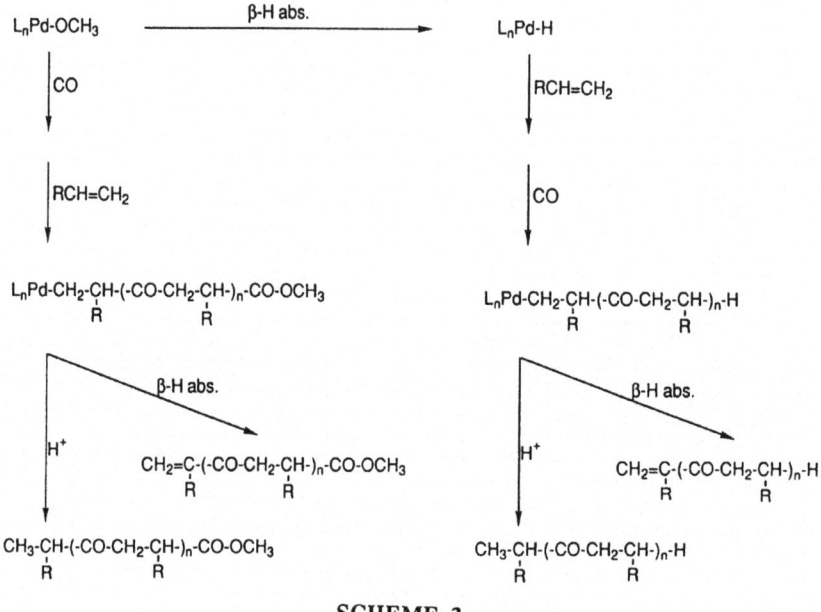

SCHEME 3

Pd-alkyl species. This was indicated by the presence of a terminal vinyl resonance at 5.82 ppm in the E-CO copolymer generated in nitromethane/THF.

In a very recent paper,[7c] Consiglio has proposed a new chain growth mechanism involving cationic Pd-carbene species in order to account for the formation of polymers with spiroketal repeating units (Scheme 4). While this mechanism was proposed specifically for the alternating copolymerization of propylene with carbon monoxide, it was implied that other olefins may copolymerize with carbon monoxide through a similar series of steps. However, several pieces of evidence appear to rule out this mechanism. Specifically: (a) we have never observed the presence of spiroketal structures in any of our alternating olefin-carbon monoxide copolymers, including the propylene-carbon monoxide copolymer, although we have carried out the copolymerizations in a variety of protic and aprotic solvents,[4,7a] (b) the proposed mechanism for the generation of the Pd-carbene required a protic solvent (specifically, an alcohol) whereas our catalyst systems were equally effective in dry nonprotic solvents, and (c) we and others (especially Brookhart[7f]) have already demonstrated the key steps in our chain growth sequence involving alternate insertions of olefin and carbon monoxide into an initial Pd-hydride bond. We have now successfully carried out the alternating copolymerization of propylene with carbon

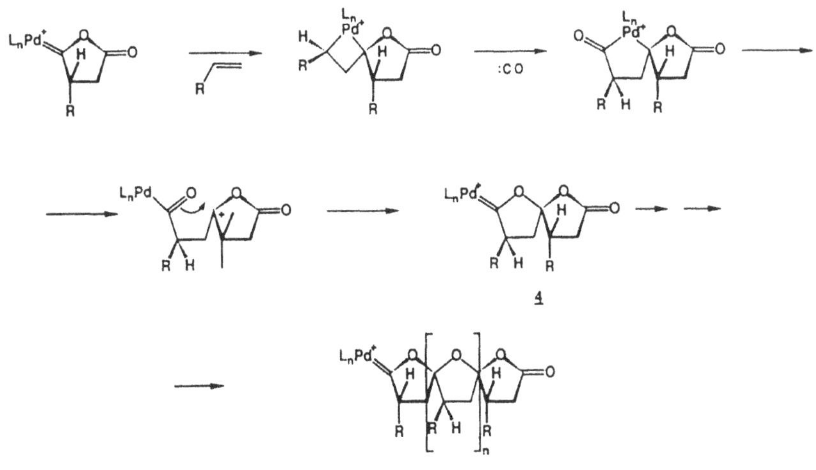

SCHEME 4

monoxide *in the absence of any solvent* by starting with the Pd-alkyl compound, [Pd(Ph$_2$P(CH$_2$)$_3$PPh$_2$)(MeCN)(Me)](BF$_4$). A copolymer with M_w = 280,000 and polydispersity (M_w/M_n) = 1.2 was obtained with a reaction rate comparable to that observed for solution-phase copolymerization using [Pd(Ph$_2$P(CH$_2$)$_3$PPh$_2$)(MeCN)$_2$]-(BF$_4$)$_2$ as catalyst. *The solid-state ^{13}C-NMR of this polymer revealed the complete absence of any spiroketal repeating units.* Based on the above observations, we believe that the Pd(II) catalyzed alternating olefin-carbon monoxide copolymerization proceeds in all case through the chain growth sequence shown in Scheme 1 and that the spiroketal formation occurs *subsequently* under certain reaction conditions. Scheme 5 outlines *one* possible series of steps leading to the formation of a spiroketal structure and the process may be assisted by the presence of Lewis acids and/or the coordination of the end carbonyl oxygen to the Pd(II) ion.

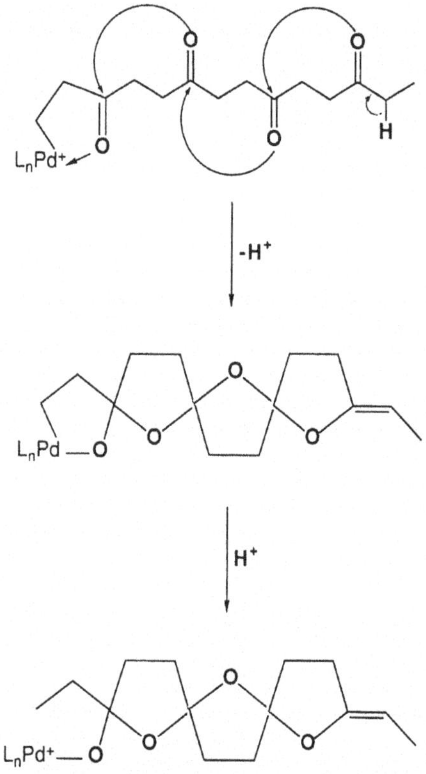

SCHEME 5

ACKNOWLEDGEMENTS

Our research was supported by a grant from the U. S. Department of Energy, Office of Basic Energy Sciences (DE-FG02-84ER13295). We also thank Johnson Matthey, Inc. for a generous loan of palladium.

REFERENCES

1. Reviews: (a) Sen, A. *Adv. Polym. Sci.*, **1986**, *73/74*, 125. (b) Sen, A. *CHEMTECH*, **1986**, 48.

2. Guillet, J. *Polymer Photophysics and Photochemistry*; Cambridge University: Cambridge, 1985; p. 261.

3. Previous reports: (a) Sen, A.; Jiang, Z.; Chen, J.-T. *Macromolecules* **1989**, *22*, 2012. (b) Jiang, Z.; Sen, A. *Macromolecules* **1992**, *25*, 880.

4. (a) Sen, A.; Lai, T.-W. *J. Am. Chem. Soc.* **1982**, *104*, 3520. (b) Lai, T.-W.; Sen, A. *Organometallics*, **1984**, *3*, 866.

5. (a) Reference 1. Recent reports: (b) Drent, E.; Van Broekhoven, J. A. M.; Doyle, M. J. *J. Organomet. Chem.* **1991**, *417*, 235.(c) Consiglio, G.; Studer, B.; Oldani, F.; Pino, P. *J. Mol. Catal.* **1990**, *58*, L9. (d) Klabunde, U.; Tulip, T. H.; Roe, D. C.; Ittel, S. D. *J. Organomet. Chem.* **1987**, *334*, 141.

6. (a) Drent, E.; Wife, R. L. *U. S. Patent* 4,970,294 (1990). (b) Van Leeuwen, P. W. N.; Roobeek, C. F.; Wong, P.K. *Eur. Pat. Appl.* EP 393,790 (1990). (c) Wong, P. K. *Eur. Pat. Appl.* EP 384,517 (1990). (d) Van Deursen, J. H.; Van Doorn, J. A.; Drent, E.; Wong, P. K. *Eur. Pat. Appl.* EP 390,237 (1990). (e) Drent, E. *Eur. Pat. Appl.* EP 390,292 (1990). (f) Drent, E. *Eur. Pat. Appl.* EP 229,408 (1986). (g) Drent, E. *U. S. Patent* 4,788,279 (1988).

7. (a) Jiang, Z.; Dahlen, G. M.; Houseknecht, K.; Sen, A. *Macromolecules* **1992,** *25,* 2999. (b) Chien, J. C. W.; Zhao, A. X.; Xu, F. *Polym. Bull.* **1992,** *28,* 315. (c) Batistini, A.; Consiglio, G. *Organometallics,* **1992,** *11,* 1766. (d) Batistini, A.; Consiglio, G.; Suter, U. W. *Angew. Chem. Int. Ed. Eng.* **1992,** *31,* 303. (e) Barsacchi, M.; Consiglio, G.; Medici, L.; Petrucci, G.; Suter, U. W. *Angew. Chem. Int. Ed. Eng.* **1991,** *30,* 989. (f) Brookhart, M.; Rix, F. C.; DeSimone, J. M.; Barborak, J. C. *J. Am. Chem. Soc.* **1992,** *114,* 5894.

8. Thomas, R. R.; Sen, A. *Inorg. Synth.* **1989,** *26,* 128; **1990,** *28,* 63.

9. Specific examples: Sen, A.; Lai, T.-W. *Inorg. Chem.* **1984,** *23,* 3257.

10. Previous report on E-P-CO terpolymer: Van Broekhoven, J. A. M. *Eur. Pat. Appl.* EP 361, 584 (1990).

11. (a) Chen, J.-T.; Sen, A. *J. Am. Chem. Soc.* **1984,** *106,* 1506. (b) Sen, A.; Chen, J.-T.; Vetter, W. M.; Whittle, R. R. *J. Am. Chem. Soc.* **1987,** *109,* 148.

12. Review on the chemistry of transition metal alkoxides: Bryndza, H. E.; Tam, W. *Chem. Rev.* **1988,** *88,* 1163. Also see: Portnoy, M.; Frolow, F.; Milstein, D. *Organometallics* **1991,** *10,* 3960.

13. Sen, A. *Acc. Chem. Res.* **1988,** *21,* 421.

14. Reviews: (a) Laine, R. M.; Wilson, R. B. In *Aspects of Homogeneous Catalysis*; Ugo, R., Ed.; D. Reidel: Dordrecht, 1984; p. 217. (b) Ford, P. C. *Acc. Chem. Res.* **1981,** *14,* 31.

FUNCTIONIZED POLYOLEFINS PREPARED BY BORANE APPROACH

T. C. Chung

Department of Materials Science and Engineering
The Pennsylvania State University
University Park, PA 16802

INTRODUCTION

Despite commercial success, the inert nature of polyolefins significantly limits their end uses, particularly, those in which adhesion or compatibility with other materials is paramount. Unfortunately, there is only very limited success in the functionalization of polyolefins by the existing processes. Polyolefins are mainly prepared by Ziegler-Natta polymerization, such catalysts are normally incapable of incorporating functional group-containing monomers because of catalyst poisoning[1]. On the other hand, post-polymerization processes suffer from other problems, such as the degradation[2] of polymer backbone. It is clear that there is a fundamental need to develop new chemistry which can address the challenge of preparing functionalized polyolefins, especially polypropylene and poly(1-butene) with controllable molecular weight and functional group concentration.

RESULTS AND DISCUSSION

In the past few years, we have been investigating a new approach for functionalizing polyolefins using the borane intermediates[3-7]. The borane containing polyolefins were prepared by both direct and post-polymerizations, which were then converted to various functional polymers under mild reaction conditions. The success of this chemistry is lied on the combination of advantages, (a) the stability of borane moiety to transition metal catalysts, (b) the solubility of borane compounds in hydrocarbon solvents (hexane and toluene) used in transition metal polymerizations, (c) the reactivity of hydroboration reaction, and (d) the versatility of borane groups, which can be transformed to a remarkable variety of functionalities[8]. Many new functionalized polyolefins with various molecular architectures have been obtained based on this chemistry. Most of them would be very difficult to obtain by other existing methods.

Direct Polymerization

The direct polymerization involves borane monomers[9-11], the α-olefin containing ω-borane, and α-olefins with Ziegler-Natta catalysts as shown in Equation 1.

Equation 1

The copolymerization between borane monomer (5-hexenyl-9-BBN) and various α-olefins, such as propylene, 1-butene and 1-octene was carried out in an inert gas atmosphere at ambient temperature using $TiCl_3 \cdot AA$ and $Al(Et)_2Cl$, an isospecific catalyst. The polymerization was started by the addition of the catalyst mixture, after aging for half hour, to a solution of the two monomers in toluene. The polymer solution was very dependent on the use of α-olefin. In 1-octene case, a homogenious purple solution was observed through the whole copolymerization reaction. On the other hand, almost immediately white precipitate could be seen in the deep purple solution in the polypropylene case, which is due to the crystalline structure with high isotactic propylene content. The copolymerization was terminated after a certain reaction time by addition of isopropanol to destroy the active metal species. Excess isopropanol was used to ensure the complete coagulation of polymer from solution. These borane containing polymers were isolated from solution by simple filtration and then washed repeatedly with isopropanol.

The resulting borane containing copolymers, poly(propylene-co-5-hexenyl-9-BBN) (PP-B) and poly(1-butene-co-5-hexenyl-9-BBN) (PB-B) containing low concentration of borane monomers (< 10 mole%), are insoluble in common organic solvents at room temperature, but soluble at high temperature. On the other hand, the copolymers of poly(1-octene-co-5-hexenyl-9-BBN) (PO-B) in all compositions are soluble in most hydrocarbon solvents at room temperature. The borane concentration can be measured by solution ^{11}B NMR technique. Only a single chemical shift at 87 ppm (vs. $BF_3 \cdot OEt_2$), corresponding to a trialkylborane, was observed in all copolymers as shown in Figure 1. The same peak existed in both polymer and mónomer showing that no detectable side reactions occurred during polymerization and work-up. By using a known amount of triethylborate (chemical shift at 19 ppm) as a reference and comparing the integrated peak areas, the borane content in the copolymer can be quantitatively determined.

The borane containing copolymers are stable for long periods of time (6 months in dry-box) or at elevated temperatures (90 °C during NMR measurement) as long as O_2 is excluded. By exposing a copolymer to air, the copolymer becomes insoluble at any temperature. The crosslinking reaction is due to free radical couplings, which are formed during the oxidation of borane groups by oxygen. In this study, the borane groups in polymers were reacted by ionic processes using $NaOH/H_2O_2$ reagents at 40 °C for 3 hours. The borane groups were completely converted to the corresponding hydroxy groups even in the PP-B and PB-B heterogeneous cases. The high surface area of borane groups is apparently due to the semicrystalline microstructure (discussed later) of the copolymers. Figure 2 (a) shows ^1H NMR spectrum of the hydroxylated polypropylene (PP-OH) with 5 mole % of hydroxy groups. The peak at 3.5 ppm is corresponding to the protons ($-CH_2-OH$) adjacent to the primary alcohol.

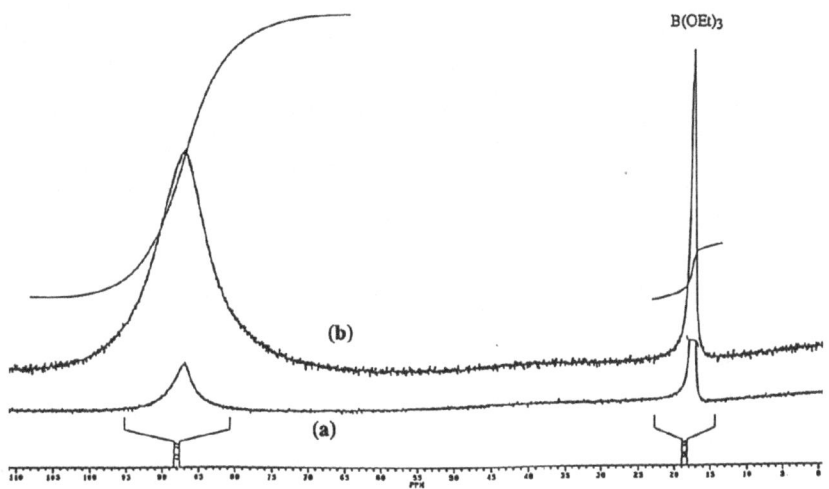

Figure 1. ^{11}B NMR spectra of (a) poly(propylene-co-hexenyl-9-BBN) with 3.5% borane mole concentration and (b) 5-hexenyl-9-BBN.

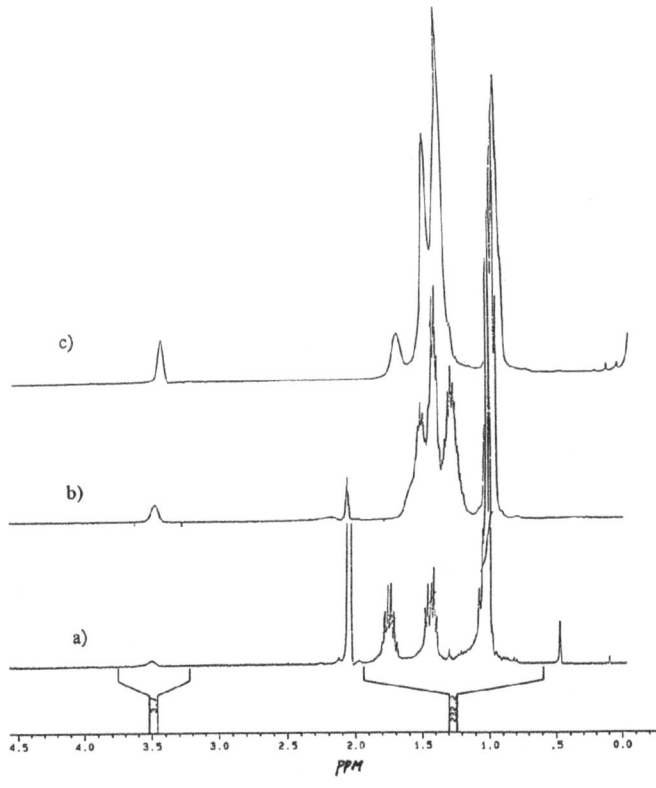

Figure 2. ^{1}H NMR spectra of (a) poly(1-propene-co-hexenol) with 5 mole% of alcohol, (b) poly(1-butene-co-hexenol) with 7 mole% of alcohol and (c) poly(1-octene-co-hexenol) 20 mole% of alcohol groups.

The PP-OH polymer can be further modified to its ester form by benzoyl chloride. The estification was complete with the disappearance of the primary alcohol peak at 3.5 ppm and the appearance of CH_2-O-C=O at 4.0 ppm and two aromatic bands around 7.2 and 8.1 ppm. To determine the copolymer compositions, the peak at 3.5 ppm in Figure 2 (a) was integrated in reference to the 3 (major) overlapping peaks between 1.95 ppm and 0.72 ppm to compare the number of protons adjacent to the alcohol (-CH_2-OH) to the rest of the aliphatic protons. Likewise, the complete oxidation reactions were shown in other polyolefin copolymers. Figure 2 (b) and (c) show hydroxylated poly(1-butene) (PB-OH) and hydroxylated poly(1-octene) (PO-OH) with 7 and 20 mole % of hydroxy groups respectively.

Borane containing polymers were also converted to the corresponding iodine containing polymers. Conversion of borane to iodine groups was complete using NaI and chloramine-T-hydrate under basic conditions at room temperature. The mild oxidizing agent chloramine-T and iodine source NaI generate I^+ in situ which reacts with the borane base complex by an S_E2 mechanism. The facile conversion is similar to those of small borane molecules[8] and borane terminated telechelic polymers[6]. The resulting iodine containing polymers were soluble in hydrocarbon solvents. However, the iodide polypropylene was initially soluble in xylene at 130°C and soon became yellowish and increasingly less soluble.

Figure 3 shows the plot of 5-hexenyl-9-BBN (mole %) in the copolymers versus reaction time.

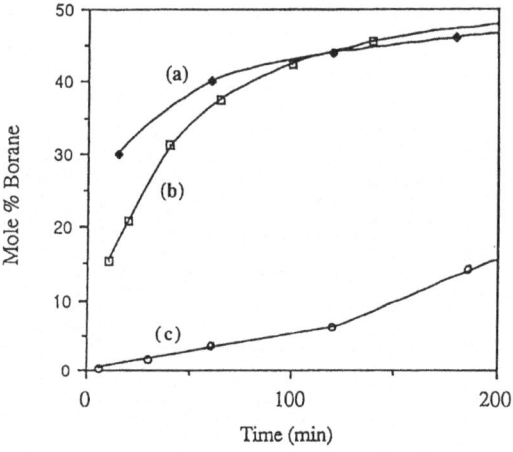

Figure 3. Plots of 5-hexenyl-9-BBN (mole %) in (a) poly(1-octene), (b) poly(1-butene) and (c) polypropylene copolymers versus reaction time.

In the copolymerization of 1-octene and 5-hexenyl-9-BBN, the copolymer compositions are quite flat and are close to the ideal 50 mole % mark as shown in Figure 3(a). Only small increase in borane content in the copolymer with the increasing conversion suggests a slightly higher reactivity of 1-octene as compared to 5-hexenyl-9-BBN. In the 1-butene case, the fluctuation of copolymer composition is much more dramatic, especially in the beginning of copolymerization. The reactivity of 1-butene is significantly higher than that of 5-hexenyl-9-BBN. In fact, the extrapolation of the composition to time = 0 in Figure 3(b), the copolymer produced at the very beginning of the reaction contains about 7 mole % 5-hexenyl-9-BBN. The initial incorporation of borane monomer in the polypropylene copolymer was very small as shown in Figure 3 (c); only 1.6 mole % in the first 0.1 hour, and then increased to 3.5 and 6 mole % after 1 and 2 hours respectively. The reactivity of the monomers is obviously different, propylene >> 1-butene > 1-octene ~ 5-hexeny-9-BBN. Basically, the reactivity of 5-hexeny-9-BBN follows the same trend in heterogeneous Ziegler-Natta polymerization, the smaller size the higher reactivity. The reactivity becomes less variable in high α-olefins.

Reactivity Ratios

The best way to describe a copolymerization is to measure the reactivity ratio of comonomers. To obtain meaningful results, a series of experiments were carried out by varying monomer feed ratio and comparing the polymer composition at low conversion. Table I compares feed and copolymer compositions at low conversions. The samples (PP-B) and (PB-B) are obtained from the copolymerization of 5-hexenyl-9-BBN/propylene and 5-hexenyl-9-BBN/1-butene respectively. The very difference in reactivity between propylene and 5-hexenyl-9-BBN significantly increases the difficulties in determining the actual amount of borane groups in copolymers using low 5-hexenyl-9-BBN concentration at low conversion. The reactivity ratio in this case gives a qualitative, rather than quantitative, expression of the monomer reactivities. On the other hand, 1-butene is much closer to 5-hexenyl-9-BBN in the reactivity, their reactivity ratios are more reliable.

Table I. A Summary of Copolymerization Between Borane Monomer and Propylene or 1-Butene at Low Conversion

Sample	Mole % B* in Feed	Mole % B* in Polymer	% Yield
PP-B	33.33	1.2	15.2
PP-B	50.00	1.7	11.0
PP-B	66.66	2.7	8.2
PB-B	14.29	2.4	9.9
PB-B	25.00	3.8	8.9
PB-B	33.33	7.0	9.4
PB-B	50.00	15.5	6.3

* B$\equiv$ 5-hexenyl-9-BBN

The reactivity ratios between α-olefin ($r_1 = k_{11}/k_{12}$) and 5-hexenyl-9-BBN ($r_2 = k_{22}/k_{21}$) are estimated by Kelen-Tudos methods[12]. The calculation is based on the equation 2.

$$\eta = r_1 \xi - r_2/\alpha (1 - \xi) \qquad \textbf{Equation 2}$$
$$\eta = G/\alpha + F \text{ and } \xi = F/\alpha + F$$

where x = [α-olefin] / [HB] in feed and y = d[α-olefin] / d[HB] mole ratio in polymer, G = x(y-1)/y, F = x^2/y, $\alpha = (F_m \times F_M)^{1/2}$. F_m and F_M are the lowest and highest values of F. Figure 5 shows the plot η versus ξ and the least squares best fit line. The extrapolation to $\xi = 0$ and $\xi = 1$ gives $-r_2/\alpha$ and r_1. We obtain $r_1 = 70.476$, $r_2 = 0.028$, and $r_1 \times r_2 = 1.973$ for propylene/5-hexenyl-9-BBN and $r_1 = 7.13$, $r_2 = 0.41$, and $r_1 \times r_2 = 2.92$ for 1-butene/5-hexenyl-9-BBN respectively. It is clear that both copolymerization reactions are not ideal cases. The values of $r_1 \times r_2$ are far from unity, and the reaction is favorable for α-olefin incorporation, especially in the copolymerization of propylene and 5-hexenyl-9-BBN. In the batch reaction with the fixed monomer ratio of propylene/5-hexenyl-9-BBN, either a broad distribution of copolymers was obtained for long reaction time or a narrow compositional distribution was obtained in short reaction time, but at extremely low yield.

Continuous Polymerization

It is feasible to obtain more uniform composition of copolymer by an engineering approach, such as the control of monomer feed ratio during the copolymerization. In the preliminary experiments, the more reactive α-olefin monomer was added gradually in order to keep its concentration constant relative to the borane monomer. The α-olefin was added in decreasing amounts to account for the consumption of borane monomer in the feed (details described in the experimental section). This approach can produce copolymer with a much narrower compositional distribution and at a higher yield (shown in Table II) of borane monomer than the corresponding one shot monomer addition in batch reactor.

Table II. The Summary of Copolymerization of α-Olefin and borane Monomer by Continuous Reaction.

Polymer	mole % B* in Feed	mole % OH in Polymer	Reaction Time (hr.)	% Yield	η**	Mv (g/mole)
PP	0	0	2	93	2.07	230,000
PP-OH	10	3	3	62	1.78	183,000
PP-OH	13	5	5	35	1.71	174,000
PB-OH	5	2.5	2	70	-	-
PB-OH	10	6.5	2	66	-	-

* B: 5-hexenyl-9-BBN

** η : intrinsic viscosity

The molecular weights of polymers were determined by intrinsic viscosity which was measured in cone/plate viscometer at 135 ^{0}C in decalin solution. To enhance the solubility of functionalized polymers, the hydroxylated polymers were completely estified with benzoyl chloride. The viscosity average molecular weights (Mv) were calculated using the Mark-Houwink equation, $[\eta]_0 = K(M_V)^a$, where K is 11.0 x 10-3 (ml/g)[13] and a is 0.80. As shown in Table II, Mv's are high, about 200,000 g/mole, for all samples. The lack of significant change in the molecular weight due to the addition of the borane monomer is quiet interesting, especially in the heterogeneous reaction condition. Obviously, no catalyst poisoning by borane group is indicated. In addition, the solubility of borane group offers the same reaction condition as α-olefin in homopolymerization.

Table III compares the fractionation results of PP-OH copolymers obtained by both processes using 1/1 monomer feed ratio and near complete monomer conversion.

Table III. Fractionation Results (weight %) of Hydroxylated Polypropylene (PP-OH)

Sample	Methanol	MEK	Heptane	Xylene	Insoluble
PP-OH-A*	none	8.5	14.2	77.7	none
PP-OH-B**	42.6	18.9	7.6	23.1	7.7

* Sample (PP-OH-A) was prepared by continuous reaction with 10/1 propylene/borane monomer mole ratio.

** Sample (PP-OH-B) was prepared by batch reaction with 1/1 propylene/borane monomer mole ratio.

Both PP-OH-A (from continuous process) and PP-OH-B (from batch process) samples were subjected to fractionation by sequential Soxhlet extractions using various solvents, e.g. methanol, 2-butanone (MEK), heptane, and xylene all under N_2. The solvents were chosen so as to separate by polarity (OH content) and crystallinity (isotacticity and PP sequence length). The PP-OH copolymers with above 60% alcohol content are soluble in MeOH. The MEK fraction was rubbery tacky material indicative of low isotacticity. Due to the low boiling point of heptane, its fraction represents polymer with intermediate tacticity or with more (and/or random distribution) hexenol units in it which reduce crystal formation. Xylene of course should dissolve all the remaining highly isotactic polymer. It is clear that the continuous reaction offers much narrow composition distribution. Most of PP-OH-B has isotactic microstructure and high crystallinity.

Post-Polymerization

Another approach to prepare borane containing polypropylene involves hydroboration reaction of unsaturated polypropylene and the subsequent oxidation reaction to convert borane group quantitatively to hydroxy groups as illustrated in Equation 3.

$$\left(CH_2\text{-}CH\right)_{98}\text{CH}_3 \quad \left(CH_2\text{-}CH\right)_2\text{CH}_2\text{-CH=CH-CH}_3$$

H-BR$_2$ | THF, 65°C

H-BR$_2$ = HB⟨⟩ or BH$_3$

$$\left(CH_2\text{-}CH\right)_{98}\text{CH}_3 \quad \left(CH_2\text{-}CH\right)_2\text{CH}_2\text{-CH}_2\text{-CH(CH}_3)\text{-BR}_2$$

NaOH | H$_2$O$_2$

$$\left(CH_2\text{-}CH\right)_{98}\text{CH}_3 \quad \left(CH_2\text{-}CH\right)_2\text{CH}_2\text{-CH}_2\text{-CH(CH}_3)\text{-OH}$$

Equation 3

The hydroboration reactions were carried out in the heterogeneous reaction conditions by suspending the powder form of poly(propylene-co-1,4-hexadiene) (PP-1,4-HD) in THF solvent. Both diborane (BH$_3$) and dialkyborane (9-borabicyclononane, 9-BBN) were used in the hydroboration reactions. It can be expected that one diborane molecule can react with up to three double bonds. Obviously, this process can cause crosslinking reactions between polymer chains. However, that did not cause any notable difference due to the heterogeneous nature. To ensure the complete reaction of the internal double bonds, the reaction was run for 5 hours at 65 °C in THF.

The borane groups in polymers were oxidized using NaOH/H$_2$O$_2$ reagents at 40 °C for 3 hours. It is remarkable to note that the complete reactions were done in a heterogeneous solution, involving a crystalline/hydrophobic polyolefin and an aqueous reagent, under mild reaction conditions. Figure 4 compares the ^{1}H NMR spectra of copolymers before and after reactions. The concentration of the unsaturated monomer units, corresponding to the chemical shift at 5.5 ppm, decreases to the limit of NMR sensitivity and shows the secondary alcohol peak appearing at 3.5 ppm. Apparently, both hydroboration and oxidation reactions were not inhibited by the insolubility of polypropylene. The effective reaction must be due to the high surface area of reacting sites. While the polypropylene segments are crystallized, the side chains containing the double bonds are expelled out to the amorphous phase which is swellable by the appropriate solvent during the reaction. In addition, the high reactivities of hydroboration reaction and the oxidation reaction certainly enhance the effeciency of functionalization.

The concentration of functional groups can be controlled by the quantity of borane reagents as well as the percentage of double bonds in polymer. The functional group is secondary due to the internal double bonds in polypropylene. Unfortunately, only internal double bonds[4] can be introduced into polyolefins by Ziegler-Natta polymerization. In the case of direct copolymerization using borane monomer, the primary functional group located at the end of the side chain was obtained in the functionalized polypropylene.

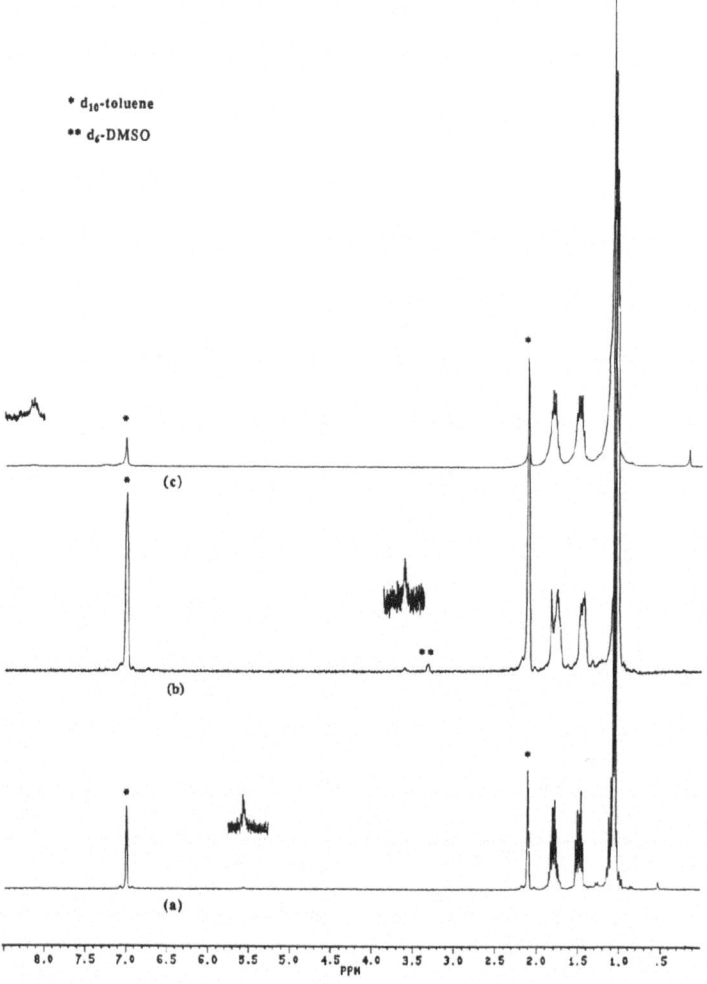

* d_{10}-toluene

** d_6-DMSO

Figure 4. The comparison of ^{1}H NMR spectra between (a) poly(propylene-co-1,4-hexadiene) with 1.7 mole % 1,4-hexadiene and its (b) hydroxylated and (c) esterified forms.

The molecular weight of polymers were determined by intrinsic viscosity which was measured in a cone/plate viscometer at 135 [0]C in decalin solution. The hydroxylated polymer (PP-OH-C) was estified with benzoyl chloride to enhance the solubility. Table IV compares polypropylene copolymers before and after functionalization.

There is no appreciable change in intrinsic viscosity and molecular weight after functionalization. The results are consistent with those of homogeneous systems, such as 1,2-polybutadiene[14], without detectable side reactions.

Physical Properties Of Hydroxylated PP

Both TGA and DSC were used to study the thermal properties of copolymers. Figure 5 compares the TGA results between two PP-OH copolymers, PP-OH-A and PP-OH-C obtained from direct and post copolymerizations respectively, and isotactic polypropylene.

Table IV. A Summary of Polypropylene Before and After Functionalization

Polymer	Intrinsic Viscosity	Mol. Weight (M_V, g/mole)	Melting Point (^0C)	Heat of Fusion (J/g)
PP-1,4-HD	1.373	131,900	153	50
PP-OH-C	1.425	138,300	151	37

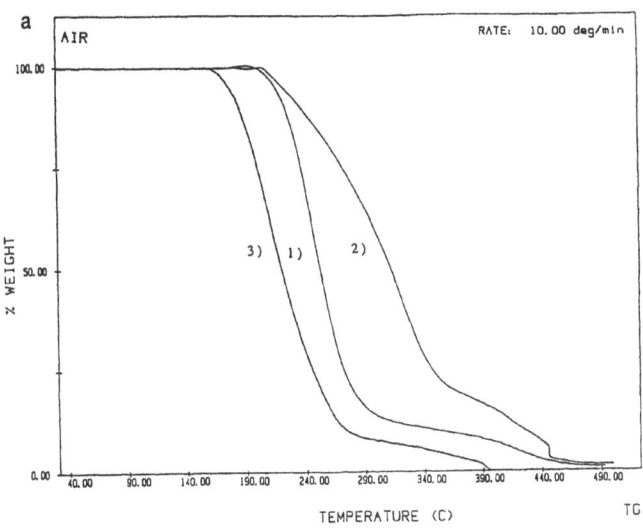

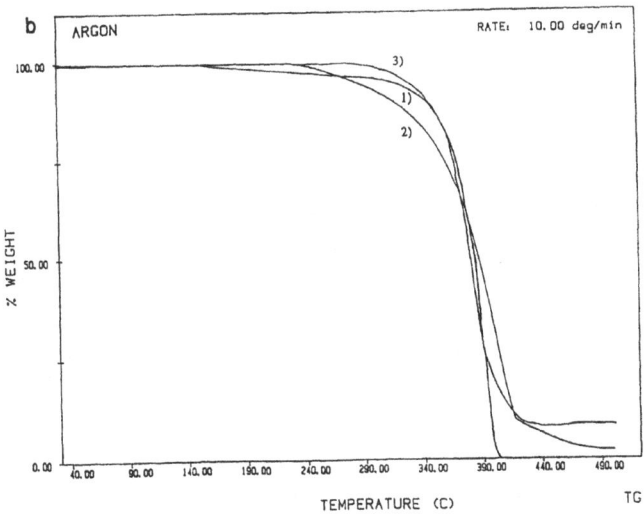

Figure 5. TGA curves of (1) isotactic polypropylene, (2) PP-OH-A with 4.2 mole% alcohol groups and (3) PP-OH-C, in air (a) and in argon (b).

The thermal stability of PP-OH-C is almost the same as that of pure isotactic PP. On the other hand, the PP-OH-A copolymer exhibits sightly better resistence in decomposition process which may be contributed from the relatively high thermal stability of the primary alcohol and the fact that the function group is pendant on a side chain. In fact, the thermal decomposition temperature of poly(vinyl alcohol) is below 150 °C. The decomposition of functional groups in the side chains does not effect the polymer backbone, therefore, the copolymers can maintain their high molecular weight and mechanical properties.

Differential Scanning Calorimetry was used to determine the effect of copolymer composition on the crystallinity. Figure 6 shows the comparison of DSC curves of polypropylene copolymers (PP-OH-A and PP-OH-C) and isotactic polypropylene. All samples have quite similar melting points (Tm ~ 160 °C), the actual depression in melting point caused by functional groups is less than 10 °C. Higher thermostability and slightly lower melting point in PP-OH copolymers ensure their processibility to form various shapes and sizes under the same conditions as i-PP.

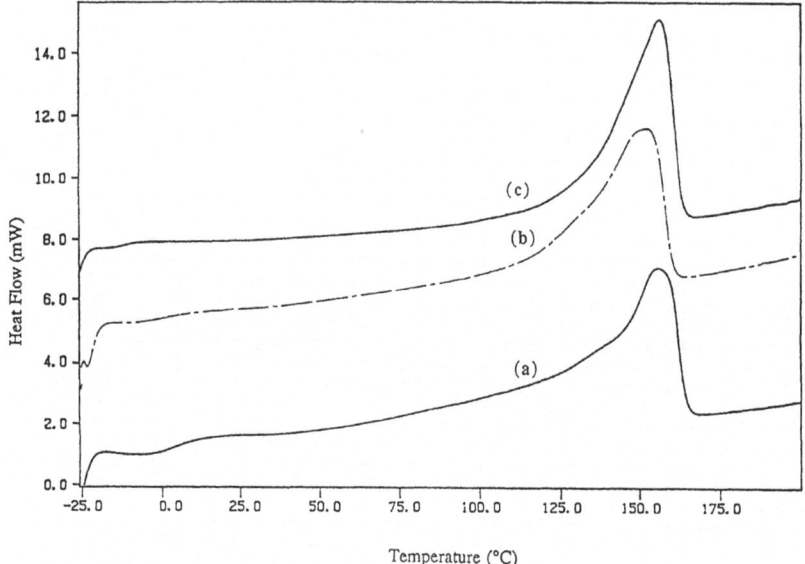

Figure 6. The comparison of DSC curves between (a) PP-OH-A, (b) PP-OH-C and (c) i-PP.

The crystalline structure of functionalized polypropylene was also studied by the polarized optical microscope. As shown in Figure 7, an isotactic polypropylene (a) was compared with two PP-OH copolymers, PP-OH-A obtained from direct copolymerization with 5 mole % of branching and the other PP-OH-C obtained by modification of unsaturated polypropylene with 2 mole % of branching. All samples were treated parallely under the same thermal condition. The spherulities are crystalline whereas the nonspherulitic regions are amorphous. It is not surprised to see the bigger spherulites in pure i-PP and smaller ones in branched PP-OH. Overall trend is consistant with DSC results, both PP-OH copolymers are highly crystalline with smaller slightly lower Tm compared with i-PP. Effective preservation of crystalline structure in the functionalzed polypropylene is apparently due to the consecutive sequences of propylene units in the polymer backbone to form the crystalline phase. Most of the hydroxy groups in polymer are located in the amorphous phases as illustrated in Figure 8. The hydroxy groups are very mobil and react with $EtAlCl_2$ almost quantitatively even in heterogeneous reaction conditions.

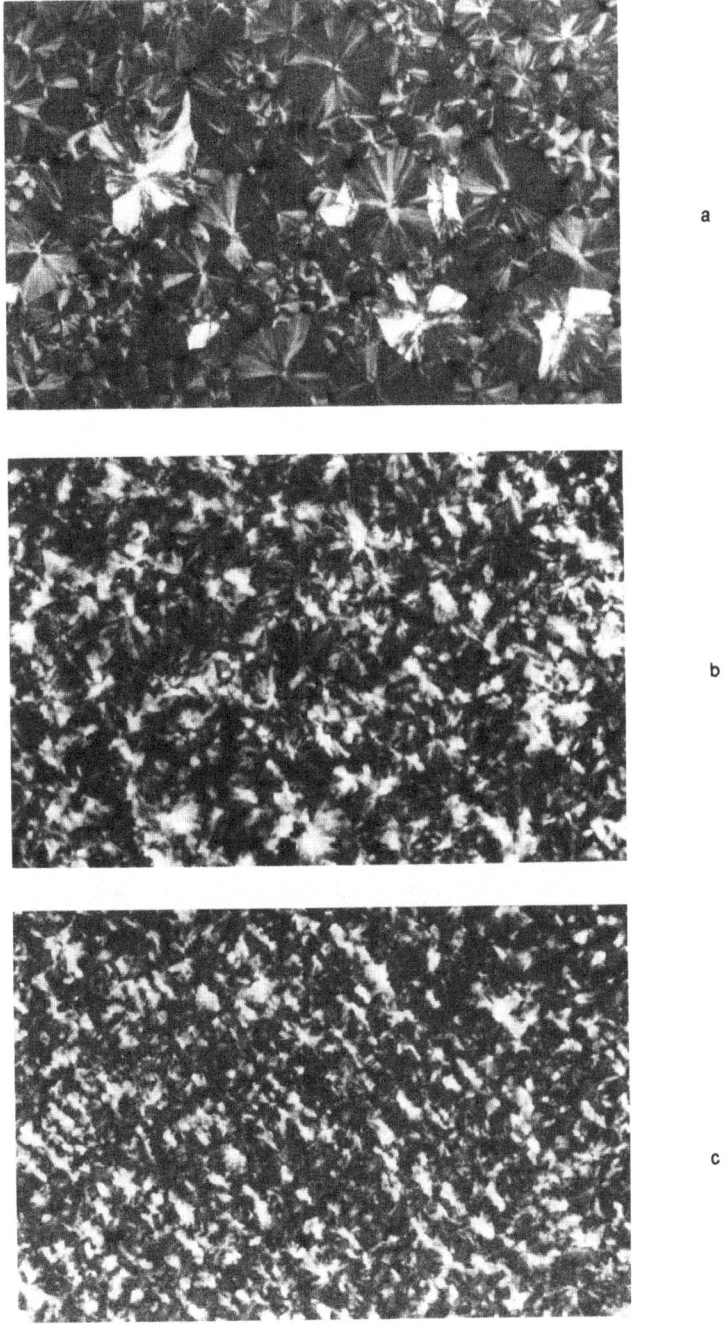

a

b

c

Figure 7. The comparison of optical micrographs between (a) isotactic polypropylene, (b) poly(propylene-co-hexe-6-ol) with 5 mole % alcohol groups and (c) poly(propylene-co-hexe-4-ol) with 1.7 mole % alcohol groups.

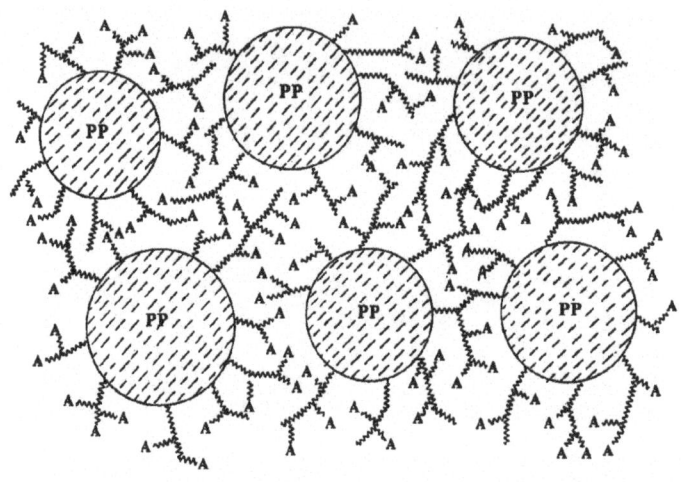

A : Functional Groups, -OH, - I, -OAlCl$_2$, -OBF$_2$

Figure 8. Pictorial description of hydroxylated polypropylene.

Adhesion Studies

The hydroxylated polypropylene was used as surface modifier to improve PP adhesion to other materials, such as aluminum (Al). Both drawn and undrawn PP films (commercial products) were laminated with PP-OH treated Al sheets in an Carver hot press. Laminates were held at 174 ^{0}C and 4 MPa for 20 min. and then slow cooled under pressure and vacuum for 5 hrs. As shown in Table V, the peel test results are compared with those obtained from standard acid etched samples which involve surface modification of both PP and Al by a dichromate-sulfuric solution and then following the same lamination procedure. PP/Al laminates bonded by PP-OH were found to exhibit an extraordinary 7-10 fold increase in peel strength over acid etched samples. Contact angles of peeled Al and PP surfaces reveal typical hydrophobic surfaces with 130 to 140^0. The same results were revealed in SEM studies. As shown in Figure 9, both peeled Al and PP surfaces show similar morphologies. It is clear that cohesive failure occurs and gives rise to the high peel strengths observed for these PP/Al laminates. When peeling Al from well bonded PP the failure path appeared to propagate within the PP layers. It also help to explain the peel strength of drawn PP/Al laminates are greater than those of undrawn PP samples.

Table V. Peel Strength of PP/Al Laminates

Sample	Peel Strength (N/m)
Acid etched	
- Undrawn PP/Al	126 ± 26
- Drawn PP/Al	130 ± 34
PP-OH Solution Cast	
- Undrawn PP/Al	675 ± 44
- Drawn PP/Al	1155 ± 52

a

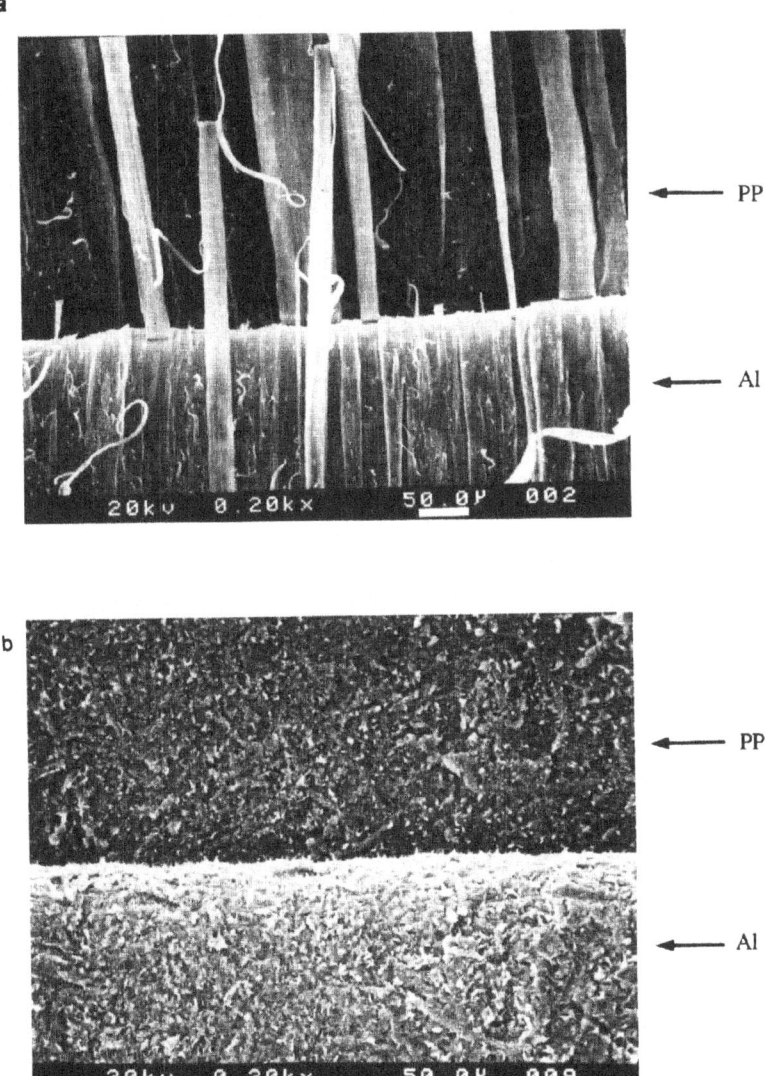

PP

Al

PP

Al

Figure 9. Micrographs of peeled surface. (a) Drawn PP/Al with extensive fibrils at the interface and (b) undrawn PP/Al without fibrils, but still exhibits cohesive failure.

The successful bonding of PP and Al by using PP-OH interfacing modifier has both scientific and commercial interests. The flexibility of hydroxy groups located at the ends of side chains in PP-OH may enhance the interaction of PP-OH and Al surface. On the other hand, high crystallinity of PP-OH (similar Tm as PP) may co-crystallize with PP, which provides strong adhesion. PP/Al laminates are potential materials of interest since such laminates should combine the drawability of PP and high modulus of Al.

EXPERIMENTAL DETAILS

Copolymerization of Propylene and Hexeny-9-BBN in Batch Reaction

A typical copolymerization, 15.5 ml (10.09 g, 0.2401 mol) of dried propylene liquid at -78 °C was then transferred into a 500 ml schlenk flask containing 200 ml of degassed frozen toluene. This reaction flask was then warmed to room temperature before bringing into the dry-box where 49.03 g (0.2401 mol) of hexenyl-9-BBN was added via the side-arm bulb. Note that the reactor contains a negative pressure. The residual monomer was washed in with 10 ml of toluene.

Meanwhile 4.093 g (5.585×10^{-2} mole) of AlEt$_2$Cl in 15 ml of toluene was added dropwise to the burgundy colored slurry of 15 ml toluene and 0.901 g (5.97×10^{-3} mol) of solid TiCl$_3$AA. This catalyst was premixed for 1/2 hour before adding to the reactor via the side-arm bulb. After 2 hours of polymerizing at room temperature cold isopropanol (20 ml) was added to terminate the polymerization as shown by the color change from deep burgundy to clear, pale brown. The reactor contents were then poured into a bottle containing 300 ml of isopropanol. The bottle was sealed and placed in the dry-box freezer overnight to facilitate the polymer precipitation or coagulation (depending on the borane content). The polymer was isolated by filtration, washed with more isopropanol, and squeeze dried, all in the dry-box. A small amount of white rubbery polymer was vacuum dried, dissolved in d$_{10}$-o-xylene, and analyzed by ^{11}B-NMR at 120 °C using triethylborate as a standard.

Copolymerization of Propylene and Hexeny-9-BBN in Continuous Reaction

In an argon filled dry box, 15.477 g of 5-hexenyl-9-BBN and 200 ml of hexane were placed in a a Parr 450 ml stirred pressure reactor and sealed. Outside the box, 4.6 g of propylene was added under N$_2$ pressure. A slurry of 1.027 g of TiCl$_3$•AA and 4.705 g of AlEt$_2$Cl in 80 ml of toluene was then added under N$_2$ pressure to catalyze the copolymerization. Additional propylene was added at 30 minute intervals with 4.10, 3.60, 3.20, 2.80 and 2.00 g of propylene added respectively. After the last monomer charge, the reaction was performed for an additional hour before terminating by injection of 100 ml of isopropanol. The reaction was stirred for additional 1/2 hour before venting the excess pressure and taking into the box for further purification with IPA.

Hydroboration Reaction

In an argon filled dry-box, 0.550 g of inhibitor-free poly(propylene-co-1,4-hexadiene) containing 1.7% of 1,4-hexadiene was placed in a suspension of 25 ml of dry, degassed toluene. A solution of 0.069 g of 9-BBN in 15 ml of dry, degassed THF was added to the polymer suspension. The suspension was heated to 65 °C in a flask equipped with a condenser. After stirring for 5 hours, the polymer was precipitated into 150 ml of dry, degassed isopropanol and isolated by filtration in the dry-box. The borane-containing polymer was then placed in a suspension of THF for oxidation reaction.

For the hydroboration reaction with BH$_3$, 0.451 g of inhibitor-free poly(propylene-co-1,4-hexadiene) containing 1.7% of 1,4-hexadiene was placed in a suspension of 15 ml dry, O$_2$-free THF in an argon filled dry-box. The polymer was hydroborated by the addition of 0.953 g of 1M BH$_3$/THF (d=0.898 g/ml) solution. The polymer slurry was heated to 65°C for 5 hours in a flask equipped with a condenser.

Oxidation and Iodozation Reactions

In a typical example, the borane containing polymer (PP-B) was placed in a 2000 ml round bottom equipped with septum and stirrer in 700 ml of THF to form an inhomogeneous slurry. A solution of 19 g of NaOH in 60 ml of water, 30 ml THF, and 20 ml MeOH was degassed and added dropwise into the reactor. The flask was then cooled to 0 °C before slowly adding a degassed solution of 87.6 g of 30% aqueous H$_2$O$_2$ in 75 ml THF via a double tipped needle. The reaction was allowed to slowly warm to room temperature before heating up to 40 °C for 6 hours. The hydroxylated polymer was then precipitated in water, squeeze dried, and placed in a slurry 500 ml of methanol. After 3 hours of vigorous stirring, approximately 75 ml of MeOH was distilled under N$_2$ to remove boric acid-methanol azeotrope. Again the polymer was precipitated in water, squeeze dried, washed with acetone, and dried under high vacuum at 45 °C.

For the iodozation reaction, 0.231 g white powdery of PP-B copolymer was placed in a suspension of THF in a 150 ml round bottom flask equipped with a magnetic stir bar and a rubber septum. A degassed solution of 0.012 g of sodium acetate in 10ml of methanol was added via a syringe under N_2. This was followed by the drop-wise addition of 0.020 g of NaI in 5 ml of degassed water. The cloudy white polymer suspension turned pale yellow and faded after 0.016 g of chloramine-T-hydrate in 10 ml of degassed MeOH was added. After 2 hours of stirring at room temperature the reaction was terminated by addition of aqueous sodium thiosulfate followed by 300 ml of dilute HCl to precipitate the polymers. The polymer was repeatedly washed with water and acetone. The solid was placed in a refluxing MeOH slurry and precipitated in water two times before drying under vacuum. The infra-red spectrum was analyzed for a KBr pellet of the white plastic.

CONCLUSION

In this paper, we have shown a new method, using borane approach, to prepare functionalized polyolefins. The processes involve both direct and post polymerizations. The success of this chemistry is lied on the combination of advantages, (a) the stability of borane moiety to transition metal catalysts, (b) the solubility of borane compounds in hydrocarbon solvents (hexane and toluene) used in transition metal polymerizations, (c) the reactivity of hydroboration reaction, and (d) the versatility of borane groups, which can be transformed to a remarkable variety of functionalities. Many new functionalized polyolefins with various molecular architectures have been obtained based on this chemistry. Most of them would be very difficult to obtain by other existing methods. The preservation of thermal stability and crystallinity in functionalized polypropylene offer the good processibility and unique morphology, in which the functional groups are located on the surface of the crystalline phases. This type of copolymer has been demonstrated as a very effective interfacial modifier to improve the adhesion between polyolefin and substrates. In addition, the copolymer is also very interesting supporting material in the application of immobilized catalysts.

ACKNOWLEDGEMENT

Authors would like to thank the financial support from the Polymer Program of the National Science Foundation.

REFERENCES

1. J. Boor, Jr., "Ziegler-Natta Catalysts and Polymerizations", Academic Press, New York (1979).
2. G, Ruggeri, M. Aglietto, A. Petragnani and F. Ciardelli, Eur. Polymer J., 19:863 (1983).
3. T. C. Chung, U. S. Patent 4,734,472 and 4,751,276 (1988).
4. T. C. Chung, Macromolecules 21:865 (1988).
5. T. C. Chung, S. Ramakrishnan and M. W. Kim, Macromolecules 24:2675 (1991).
6. T. C. Chung and M. Chasmawala, Macromolecules, 24:3718 (1991).
7. S. Ramakrishnan and T. C. Chung, Macromolecules, 23:4519 (1990).
8. H. C. Brown, "Organic Synthesis via Boranes", Wiley-Interscience, New York (1975).
9. T. C. Chung and D. Rhubright, Macromolecules 24:970 (1991).
10. T. C. Chung, ChemTech 21:496 (1991).
11. S. Ramakrishnan, E. Berluche and T. C. Chung, Macromolecules 23:378 (1990).
12. T. Kelen and F. Tudos, React. Kinet. Catal. Lett., 1:487 (1974).
13. J. B. Kinsinger and R. E. Higher, J. Phys. Chem., 63:2002 (1959).
14. T. C. Chung, M. Raate, E. Berluche and D. N. Schulz, Macromolecules, 21:1903 (1988).

SYNTHESIS OF POLYOLEFIN-PMMA GRAFT COPOLYMERS

T.C. Chung*, G.J. Jiang, and D.C. Rhubright

Department of Materials Science and Engineering
The Pennsylvania State University
University Park, PA 16802

INTRODUCTION

Due to low surface energy, lack of chemical functionalities, and crystallinity polyolefins possess poor interactions with other materials. Polyolefins exhibit inadequate compatibility with other synthetic polymers and virtually no adhesion to metal or glass. If these problems could be overcome it would phenomenally expand the available market for polyolefin applications. Accordingly, the chemical modification of polyolefins, especially polypropylene, has been an area of intense interest as a route to improve these commodity polymers.

An established technique for improving the interfacial interaction between polyolefins and other materials is the use of graft and block copolymers as compatibilizers[1-3]. It is very desirable to prepare polyolefin graft copolymers containing functional polymers, such as PMMA, PVA, at the side chains which can dramatically increase the interaction of polyolefin with a broad range of polymers containing functional groups. Unfortunately, the chemistry to prepare polyolefin graft copolymers and functionalized polyolefins are very limited, mainly due to the inert nature of polymers and catalyst poison by functional groups.

Numerous methods have been employed in forming graft copolymers with polyolefins. Ionizing radiation (x-ray, γ-rays, and e-beams) in the presence of air, ozone, uv with accelerators, and free radical initiators have all been used to form polymeric peroxides[4,5]. When heated in the presence of monomers, the polymeric peroxides can initiate graft polymerizations. Typically, these high energy reactions lead to side reactions such as crosslinking and chain cleavage resulting in diminished mechanical properties. In most cases, the structure and composition of copolymers are difficult to controlled with the considerable amounts of ungrafted homopolymers.

RESULTS AND DISCUSSION

This paper describes a new general method to prepare polyolefin graft copolymers which includes most polyolefin backbones and a wide selection of free radically polymerized grafts. Our method involves borane-containing polyolefins and free radical graft-from polymerization as illustrated in Equation 1. The oxidation reaction of borane groups results in free radicals pendant from the polyolefin backbone, such as poly(1-octene) and polypropylene. The polymeric free radicals initiate the graft-from polymerization of methylmethacrylate (MMA). This reaction occurs at room temperature with no chain degradation.

---

* Author to whom all correspondence should be addressed.

New Advances in Polyolefins, Edited by
T.C. Chung, Plenum Press, New York, 1993

Equation 1

$$-(CH_2-CH)_x-(CH_2-CH)_y- \xrightarrow[\text{room temp}]{MMA/O_2} -(CH_2-CH)_x-(CH_2-CH)_y-$$

with R and (CH$_2$)$_4$–B substituents on the left, and R and (CH$_2$)$_4$–PMMA substituents on the right

Equation 1

To achieve high graft efficiency, high yield of graft copolymers with minimizing the PMMA homopolymer formation, it is essential to control the oxidation reaction of trialkyborane at the primary C-B bond attached to the polyolefin, instead of the secondary C-B bonds of the cyclooctyl ring. The selective reaction will be discussed in the next section.

Selective Autoxidation of Alkyl-9-BBN

In general, the autoxidation of boranes proceeds through a free radical homolytic chain mechanism[6,7]. The driving force is the conversion of B-C bonds (107 kcal/mol) to the stronger B-O bonds (192 kcal/mol). Oxygen donates e$^-$ density to stabilize the e$^-$ deficient boron, whereas carbon cannot backbond in trialkylboranes. Oxygen will oxidize boranes producing boron peroxides which can react further with an alkyl borane as shown in Equation 2. The reduction of the peroxide by another trialkylborane yields an alkyl radical and a borinate B-O• radical which is relatively stable and is inactive for polymerization. The boron peroxides will also homolytically cleave to generate an alkoxy radical and a B-O• radical. The alkyl and alkoxy radicals can then initiate the polymerization of methacrylates, styrene, acrylamide, vinyl acetate, acrylonitrile, etc. at room temperature.

$$R-CH_2-B \diagup + O_2 \longrightarrow R-CH_2-O-O-B \diagup \longrightarrow \; \bullet O-B \diagup + \; R-CH_2-O\bullet$$

$$\downarrow R'-CH_2-B\diagup$$

$$R-CH_2-O-B\diagup + \; \bullet O-B\diagup + \; R'-CH_2\bullet$$

Equation 2

Theoretically, if the alkyl groups on boron had equal reactivity towards O$_2$, then only 1/3 of MMA produced would be graft and 2/3 would be MMA homopolymer for trialkylborane initiator with its 3 B-C bonds. As will be shown in the next section, the PMMA generated is almost exclusively graft copolymer covalently attached to the polyolefin. This improved grafting is attributed to the higher reactivity of the unhindered polymeric primary C-B bond as opposed to the more sterically hindered secondary C-B bonds of the cyclooctyl ring. Oxygen insertion into a cyclooctyl B-C bond would need to distort the cyclooctyl's stable double chair conformation in order to form the boron peroxide.

Synthesis of Poly(1-octene)-g-PMMA Copolymers

The borane grafting method is first demonstrated in the synthesis of poly(1-octene)-g-PMMA.[8] Poly(1-octene) was chosen as the backbone because of its ease of preparation as a borane-containing copolymer and its excellent elastomeric properties, with a T_g of -65 °C and an extremely low plateau modulus. In addition, the completely saturated polymer structure offers excellent chemical and thermal stability; its decomposition temperature is above 350 °C in air. The rubbery poly(1-octene) backbone with grafts of glassy PMMA (T_g=112 °C) behaves as a thermoplastic elastomer.

The poly(1-octene) containing pendant alkyl-9-BBN was prepared by copolymerizing B-5-hexenyl-9-BBN and 1-octene with a Ziegler-Natta catalyst.[9] In this study, the copolymer contained 0.3 mole % of B-5-hexenyl-9-BBN was prepared by using monomer feed ratio of 1-octene/hexenyl-9-BBN = 200/1 and 1 hour reaction time. The copolymer yield is close to 60 %. The graft-from reaction was carried out at room temperature by mixing poly(1-octene-co-hexenyl-9-BBN) with MMA monomers in neat or with solvent, such as THF. Although the stoichiometry amount of oxygen to boron is needed in the oxidation reaction, the best

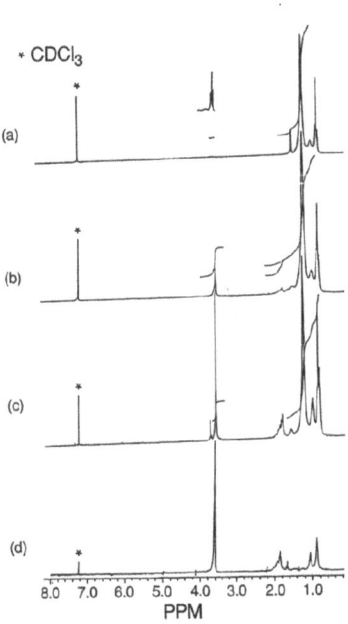

FIGURE 1. ^{1}H NMR of poly(octene-*co*-hexenol) with 0.3 mol % hydroxy groups (a), poly(1-octene)-g-PMMA with 21 (b) and 65 (c) mol % MMA, and PMMA homopolymer (d).

results are realized when the O_2 is introduced slowly so that O « B at any time. Excess O_2 is a poison for free radical polymerizations. Also, high O_2 concentrations lead to over oxidation to boronates and borates which are poor free radical initiators at room temperature. In a typical graft-from reaction, only about 10% of oxygen was added in each time. A notable increase in solution viscosity was observed within few minutes after injecting. The reaction mixture was continued at ambient temperature for few hours before exposing the reaction mixture to air which terminated the reaction. The resulting product usually shows a significant weight increase from poly(1-octene-co-hexenyl-9-BBN).

To remove PMMA and poly(1-octene) homopolymers, solvent extractions by acetone and hexane were carried out by refluxing each solvent through the product for 24 hours. Only a negligible amount of acetone and hexane soluble fractions were obtained. The acetone and hexane insoluble polymer is soluble in THF and is poly(1-octene-g-MMA). To assure the efficiency of seperation, a control experiment was also carried out by mixing two homopolymers, poly(1-octene) and PMMA with equal amount, in THF. This polymer mixture was then precipitated in isopropanol, and was subjected to the same extraction condition using refluxing acetone and hexane. Most of PMMA homopolymer was recovered in acetone-soluble fraction with no detectable amount of PMMA left in poly(1-octene) which is soluble in hexane.

Figure 1 compares the ^{1}H NMR spectra of poly(octene-g-MMA) copolymers, poly(1-octene-co-hexenol), and PMMA. The pure PMMA homopolymer was obtained by a control reaction using a model compound hexyl-9-BBN as an initiator under similar polymerization conditions. In Figure 1a, a small triplet peak at 3.5 ppm, corresponding to the primary alcohol methylene group (CH_2OH), indicates 0.3 mol % of hexenol in poly(1-octene-co-hexenol). The chemical shifts at 3.6 and 1.8 ppm in Figure 1b,c correspond to methyl groups (CH_3O) and methylene groups, respectively, in PMMA. The chemical shifts between 1.6 and 0.7 ppm include all protons in polyoctene and three protons in the methyl group located on the PMMA backbone. The copolymer composition was calculated by the ratio of the two integrated intensities between 3.6 and 1.6 - 0.7 ppm and the number of protons both chemical shifts represent. Spectra b and c of Figure 1 indicate 21 and 65 mol % of PMMA, respectively, in poly(1-octene)-g-PMMA copolymers.

In a GPC study only a single peak at high molecular weight was observed in the graft copolymer, with no peak for the PMMA homopolymer which is expected to have a molecular weight of about 65,000 g/mole. Unfortunately, the molecular weight of the graft copolymer is difficult to measure by simple GPC with an RI detector, especially in this case, which involves high molecular weight ($>1.5 \times 10^6$ g/mole) poly(1-octene-co-hexenyl-9-BBN) as the backbone polymer. The average chain length of PMMA can be estimated by the NMR determination of PMMA and the moles of borane groups in poly(1-octene-co-hexenyl-9-BBN). Assuming all organoborane groups (0.3 mol %) are involved in the graft-from reaction, the average molecular weight of PMMA in the resulting poly(1-octene-g-MMA) with 65 mol % of PMMA is about $Mn \approx 61,000$ g/mole. In a control reaction using similar reaction conditions and hexyl-9-BBN initiator, the PMMA homopolymer has a similar molecular weight, $Mn = 65,000$ g/mole. It is assumed that each PMMA chain has a similar molecular weight despite the reaction time. The prolongation in reaction time increases the oxidation reaction of the borane group and increases the graft density. Unreacted boranes are eventually oxidized to OH's by O_2 and H_2O. The final graft density is dependent on the borane concentration in poly(1-octene-co-B-5-hexenyl-9-BBN) copolymer.

Synthesis of PP-g-PMMA Copolymers

The other polyolefin used to demonstrate the borane approach is polypropylene. Polypropylene was chosen due to its excellent properties as a highly cystalline thermoplastic. Traditionally, i-PP has proven difficult to controllably graft to yield well-defined structures.

Borane group containing polypropylene copolymers can be obtained from both direct and post-polymerizations. In the direct polymerization, copolymer was synthesized using a combination of isospecific catalyst TiCl3•AA/AlEt2Cl and propylene/borane monomer (5-hexenyl-9-BBN)[11]. As previously demonstrated[10], Ziegler-Natta catalysts are stable in the presence of Lewis acid boron alkyls. Therefore, borane monomer can copolymerize with propylene unlike other functional monomers which poison the metal-alkyl catalyst. The ^{11}B NMR of copolymer (I) showed a single peak at 87 ppm, the same as the shift for the unpolymerized monomer, indicating an unaltered trialkylborane. In this study, the copolymer containing 0.5 mole% of borane monomers was synthesized with high molecular weight (Mv ~200,000 g/mole). The molecular weight was determined by intrinstic viscosity of the corresponding hydroxylated polypropylene which was obained by oxidation reaction of borane containing polypropylene (I) with $NaOH/H_2O_2$.

Alternatively, borane-containing polypropylene (II) was also prepared by hydroboration of commercially available unsaturated polypropylene[12], such as propylene and 1,4-hexadiene copolymer. The internal double bond on the hexadiene side chain was reacted by 9-BBN (9-borabicyclonornane). The reaction is very effective despite the heterogenious condition. The concentration of borane groups was controlled by the amount of 9-BBN used in the hydroboration reaction and the concentration of unsaturation (usually less than 2 mole%) in the unsaturated PP copolymers.

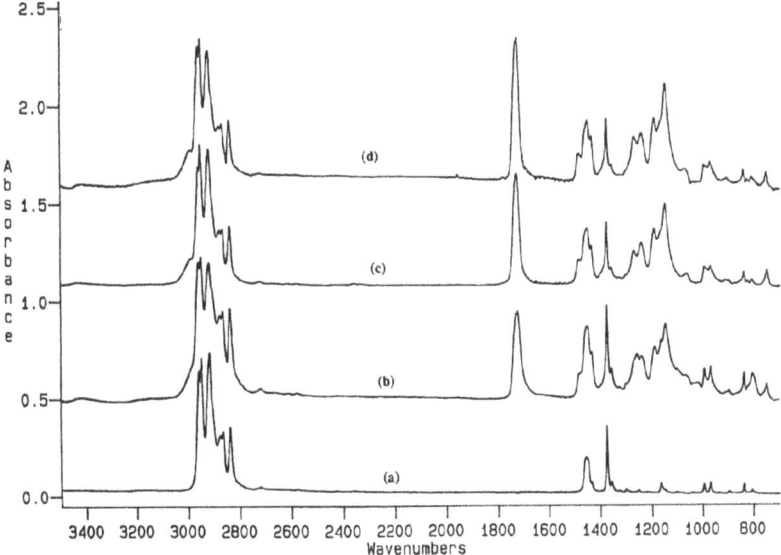

The borane groups are primany in direct polymerized PP (I) and secondary in post-polymerized PP (II). Both are extremely air-sensitive, are stable indefinitely when sealed under vacuum or as long as one year in a dry-box. Also, samples were stable for short periods at elevated temperatures, up to 120°C during NMR measurements. By exposing the suspended borane-containing PP in uninhibited MMA to oxygen, the graft-from reaction was initiated in the heterogenious condition. Even though the final B:O ratio was 1:1 only 5% of the oxygen was added hourly. After 12 hours the reaction was terminated by recovering the unreacted MMA under vacuum. To remove the by-product of boric acid, the solid was refluxed in 100 ml of MeOH before distilling off 20 ml and isolated by filtration. The white solid was extracted with acetone (which can effectively separate a mixture of i-PP and MMA homopolymers) in a Soxhlet apparatus for 24 hours. Only very low% of acetone soluble fraction was recovered, which is PMMA homopolymer with an Mn ≈ 70,000 g/mole. The acetone insoluble fraction was PP-g-PMMA copolymer and was completely soluble in xylene at elevated temperatures.

FIGURE 2. FTIR of poly(propylene-*co*-hexenol) with 0.5 mol % hydoxy groups (a) and PP-*graft*-PMMA samples containing 18 (b), 52 (c) and 66 (d) mol % of MMA.

Figure 2 shows the IR spectra of the resulting PP-g-PMMA copolymers which were prepared from borane containing PP (I). The absorption band at 1730 cm-1, corresponding to ester groups, clearly indicates the existance of PMMA in the graft copolymers. The quantitative compositions were determined by ^{1}H NMR spectrum (shown in Figure 3) which were examined in d$_{10}$-o-xylene at 120°C. The chemical shifts at 3.6 ppm and 1.8 ppm correspond to methyl groups (CH$_3$-O) and methylene groups respectively in PMMA. The chemical shifts at 1.9, 1.6 and 1.1 ppm correspond to methine, methylene and methy groups in polypropylene. The quantitative analysis of copolymer composition was calculated by the ratio of two integrated intensities between δ=3.6 ppm and δ=2.1 to 1.1 ppm and the number of protons both chemical shifts represent. Figure 3 (a), (b) and (c) indicate 18, 52 and 66 mole% of PMMA respectively in PP-g-PMMA copolymers.

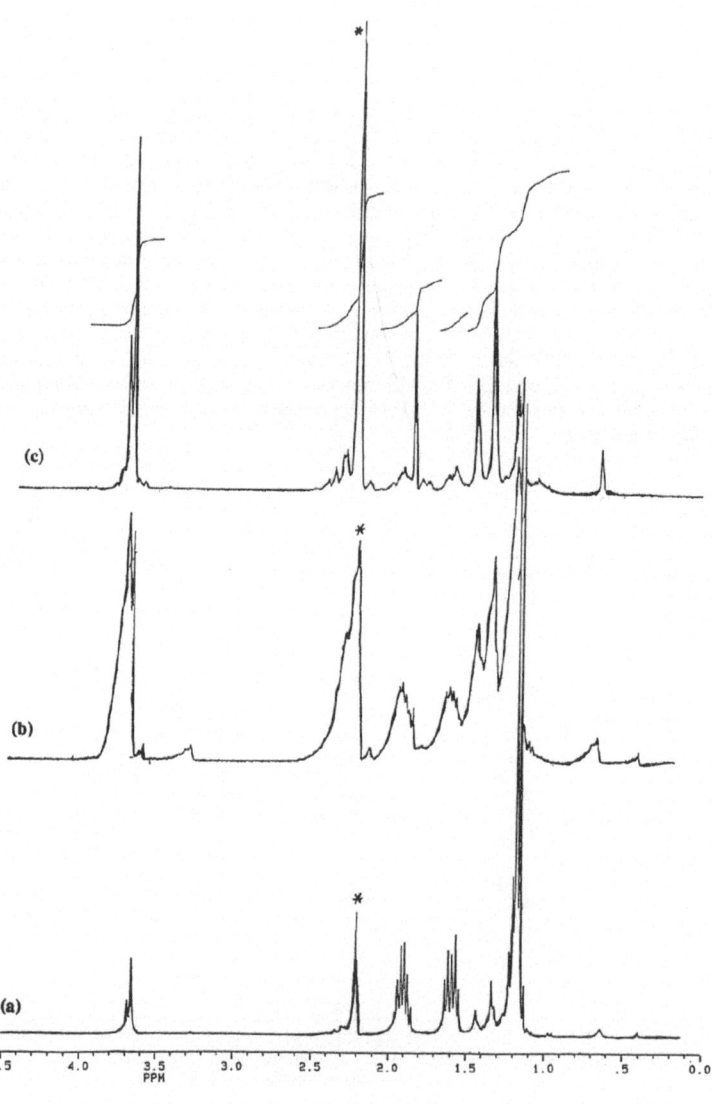

FIGURE 3. ^{1}H NMR spectrum of PP-g-PMMA samples containing 18 (a), 52 (b) and 66 (c) mol % of MMA.

Table I summarizes the experimental results of PP-g-PMMA copolymers which are obtained from both polymer (I) and (II) containing primary and secondary alkyl-9-BBN respectively. The comparison of runs 1, 2, 3 and 4 shows the sensitivity of oxygen addition to the graft efficiency. Even though the final stoichiometry of oxygen to boron should be 1:1, the best results in this heterogeneous reaction system are realized when the O_2 is introduced slowly so that O « B at any time. Excess O_2 is not only a poison for free radical polymerizations but also leads to over oxidation to boronates and borates which are poor free radical initiators at room temperature. The polarity of solution also effect the graft reaction. THF is a very good solvent in this reaction. The nopolar solvent, such as benzene, slows down the graft-from reaction, which may be due to the solubility of O_2 in the solvent. In the run 5 the oxygen was introduced by diffusion of air through the rubber septum which was tightly installed on the top of the reactor. The insufficient O_2 in this process leads to low % of PMMA formation.

TABLE 1. A Summary of PP-g-PMMA Copolymers.

Run	Mole % B in PP	O_2 ml / hr	monomer/ solvent	rxn time (hrs)	mole % MMA in polymer
1	0.5	1.5/12	MMA	48	66
2	0.5	3.0/1	MMA	2	6
3	0.5	1.4/3	THF	12	52
4	0.5	6 all at once	MMA	48	1.5
5	0.5	diffusion	THF	48	12
6 *	1.7	diffusion	MMA	24	18
7 *	1.7	1/12	benzene	12	13

* prepared from unsaturated PP

It is very interesting to note that a significant difference were observed by using primary and secondary borane groups in graft-from reactions. Despite higher borane concentration in polymer (II), polymer (I) shows higher concentration of PMMA grafted to PP. In addition, a significant high concentration (> 30%) of PMMA homopolymer was isolated by using polymer (II). High grafting efficiency in polymer (I) must be attributed to the higher reactivity of the unhindered polymeric primary C-B bond as opposed to the more sterically hindered secondary C-B bonds.

Thermal Properties

The thermal properties of the copolymer were determined by DSC measurement. The samples were first heated to 170°C and then rapidly cooled to -150°C. As shown in Figure 4, two glass transition temperatures (T_g), -63 and +113°C, exist in poly(1-octene)-g-PMMA copolymer with 50/50 composition. Both T_g's are the same as those of the two corresponding homopolymers. This indicates clear phase separation between the polyoctene backbone and the PMMA side chains. The control of graft density in the copolymer offers enough consecutive sequences of octene units in the polymer backbone to form separate domains. In fact, the poly(1-octene)-g-PMMA copolymers with about 30 % of PMMA behave like the thermoplastic elastomer with strong mechanic properties.

Figure 5 compares DSC traces of pure i-PP (upper curve, ΔH=62 J/g), PP-g-PMMA with 5% MMA (middle curve, ΔH=52 J/g), and PP-g-PMMA with 67% MMA (lower curve, ΔH=22 J/g). The melting point is almost unchanged despite high percentage of PMMA in the copolymer. The polypropylene segments in both graft copolymers perserve their crystallinity, the heat of fusion per gram of PP is almost the same in three samples. In addtion, the glass transitions for the graft copolymers are only slightly higher than the Tg of pure PP indicating that the PP and PMMA segments are phase separated. Overall, the branch density in the graft copolymer must be relatively low and polypropylene segments have enough consecutive sequences to form crystalline domains as pure PP.

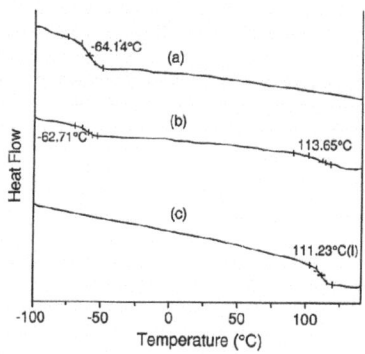

FIGURE 4. DSC curves of poly(octene-*co*-hexenol) with 0.5 mol % hydroxy groups (a), poly(1-octene)-g-PMMA with 50 mol % MMA (b), and PMMA homopolymer (c). **(b)** DSC curves of i-PP (i) and PP-*graft*-PMMA containing 5 (ii) and 66 (iii) mol % MMA.

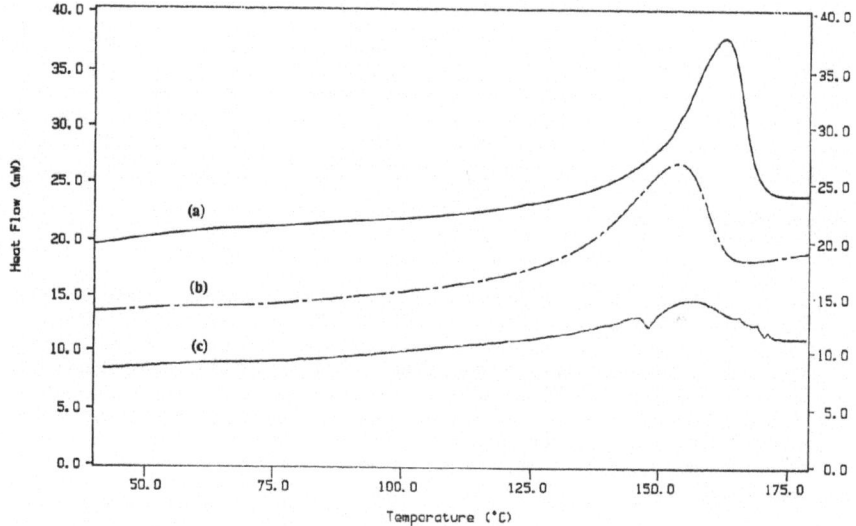

FIGURE 5. DSC curves of i-PP (a) and PP-g-PMMA containing 5 (b) and 66 (c) mol % MMA.

Thermal gravimetric analysis demonstrated that the graft copolymers begin weight loss at 190°C in air and at 250°C in argon whereas PP begins weight loss at 230°C in air and at 380°C in argon. The thermal decomposition of graft copolymer obviously starts from the less thermally stable PMMA segments.

PP/PMMA Blends

Optical microscopy was used to evaluate the graft copolymers ability to act as a phase compatibilizer for blends of PP and PMMA homopolymers in Figure 6. Polymer solutions were prepared in BHT inhibited *o*-xylene at 135°C. The polymer films were then solution cast onto glass microscope slides. After evaporating the xylene under N_2 purge, the films were covered with a slide cover. The polymer films were then melted in a hot stage at 180°C for 15 minutes. The samples were then allowed to cool quiescently in the hot stage to room temperature for 20 minutes. The magnification on the microscope was 100 times. Blend A is 70/30 weight % mixture of PP and PMMA homopolymers that were blended in solution.

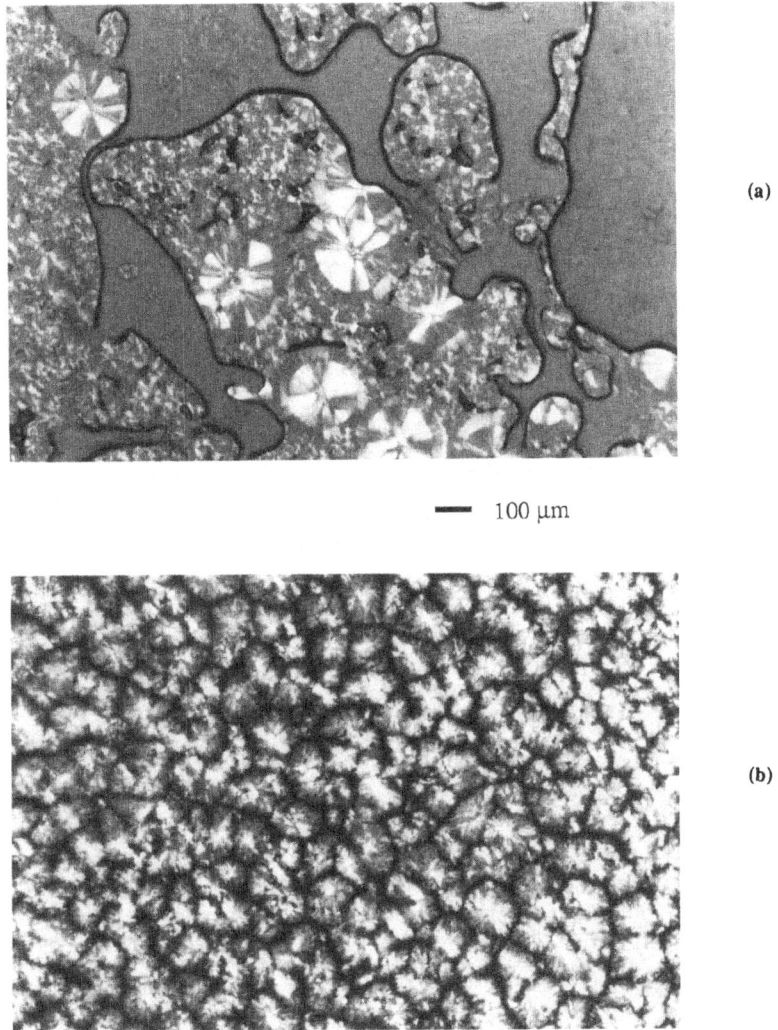

(a)

— 100 μm

(b)

FIGURE 6. Optical micrographs of polymer blend of 70/30 weight % PP/PMMA homopolymers (a) and blend of 70/30/10 of PP/PMMA/PP-g-PMMA where the graft copolymer also contains 30 % PMMA.

Two distinct phases are visible, the crystalline PP phase and an amorphous PMMA phase. Within the PP domain, the spherulite size varies greatly with a few extremely large crystllites and predominantly small spherulites. Blend B is a sample of blend A to which 10 weight % of PP-g-PMMA has been added in solution. The added graft copolymer was also 30 weight % PMMA so as not to change the overall composition. The graft copolymer behaving as a polymeric emulsifier increases the interfacial interaction between the PP amorphous region and the PMMA to reduce the domain sizes. The most noticeable change in the micrographs is the disappearance of the visibly distinct PMMA domains. The large phase seperated PMMA domains are now dispersed into the inter-spherulite regions and cannot be resolved by the resolution of the optical microscope. The mode of nucleation within the polypropylene crystalline phase has changed as evidenced by the now relatively homogeneous sperulite size. Currently, the mechanical properties of these blends are under investigation.

It should also be noted that both films of the compatibilized blend and the pure graft copolymer formed in a melt press were optically clear. This is unlike pure i-PP which forms hazy, translucent films. The lack of large spherulites in both the blend and the graft must minimize scattering.

EXPERIMENTAL DETAILS

The borane containing polymers, both polyoctene and polypropylene, were prepared by the published method[10-12]. In a typical graft-from reaction by borane approach , 2 g of poly(1-octene-co-B-5-hexenyl-9-BBN) containing 0.3 mole % organoborane was dissolved in 50 ml of THF and then mixed with 10 ml of MMA in a 200 ml flask in an argon filled dry box. The graft reaction was initiated by introduction of 5 ml of air into the septa sealed flask via a syringe. With constant stirring, a notable increase in the solution viscosity was observed within minutes. The reaction was allowed to continue at ambient temperature while the solution became highly viscous within 1 hour. After 4 hours, the reaction was terminated by exposing the reaction to the atmosphere. The product was precipitated into isopropanol, collected by filtration, and finally washed with hot isopropanol 3 times. To remove possible PMMA homopolymer from the sample, it was extracted with hot acetone in a Soxhlet apparatus for 24 hours. Only a negligible amount of acetone soluble PMMA was isolated. The acetone insoluble fraction was 3.4 g of poly(1-octene)-g-PMMA after drying. To ensure the efficiency of our separation technique, a control was carried out mixing equal amounts of the two homopolymers, poly(1-octene) and PMMA, in THF. This polymer mixture was then precipitated into isopropanol and was subjected to the same extraction procedure using refluxing acetone. PMMA homopolymer was recovered in the acetone-soluble fraction, with no detectable amount of PMMA left in the acetone insoluble poly(1-octene).

In a graft-from reaction of polypropylene, 2 g of a borane-containing copolymer prepared by direct polymerization, containing 3 mole% of borane monomer, was placed in a suspension of 12.0 g dry, degassed, and uninhibited MMA in a sealed, opaque flask. Benzene and THF were also effective media for the graft reaction although the polymer does not dissolve. The reaction was initiated by injecting dry O_2 into the reaction flask. Even though the final B:O ratio was 1:1 only 5% of the oxygen was added hourly. After 12 hours the reaction was terminated by recovering the unreacted MMA under vacuum. The polymer was refluxed in 100 ml of MeOH before distilling off 20 ml and isolating by filtration. The white solid was extracted with acetone (which can effectively separate a mixture of iPP and MMA homopolymers) in a Soxhlet apparatus for 24 hours to typically yield 95% grafted MMA and 5% homopolymer. The 3.41 g of acetone insoluble in polymer was completely soluble in xylene at elevated temperatures. ^1H NMR spectroscopy in d_{10}-o-xylene at 120°C showed that the graft copolymer contained 52 mole % MMA. The 0.08 g acetone soluble polymer was PMMA with an Mn ≈ 70,000 g/mole.

CONCLUSION

We have demonstrated the success of the borane approach to form MMA grafts on polyoctene and PP. The borane graft technique allows us to overcome the difficulties usually associated with functionalizing polyolefins. By judicious choice of α-olefin, borane monomer, and reaction conditions we can control the amount of borane monomer incorporation and the morphology of the backbone polymer. Overall, the chemistry is very general and can be applied to a broad range of polyolefin graft copolymers. In addition, grafted polyolefins will be used as thermoplastic elastomers, adhesives, rubber toughened thermoplastics, and compatiblizers in blends and composites.

ACKNOWLEDGEMENT

Authors would like to thank the financial support from the Polymer Program of the National Science Foundation.

REFERENCES

1. G. Riess, J. Periard, A. Bonderet, *Colloidal and Morphological Behavior of Block and Graft Copolymers*, Plenum: NY (1971).
2. B. Epstein, U. S. Patents 4,174,358, (1979).
3. D. Lohse, S. Datta, E. Kresge, *Macromolecules* **24**, 561 (1991).
4. G. Natta, E. Beati, and F. Severine, *J. Polymer Sci.* **34**, 548 (1959).
5. B. Ranby and F. Guo, *Polymer Preprints* **31**, 446 (1990).
6. B. Mikhailov and Y. Bubnov, *Organoboron Compounds in Organic Synthesis*, (1984)
7. H.C. Brown, *Organic Synthesis Via Boranes*; Wiley-Interscience: NY (1975).
8. T.C. Chung and G.J. Jiang, *Macromolecules* **25**, 4816 (1992).
9. T.C. Chung, U. S. Patents 4,734,472 and 4,751,276 (1988).
10. T.C. Chung, *Macromolecules* **21**, 865 (1988).
11. T.C. Chung and D. Rhubright, *Macromolecules* **24**, 970 (1991).
12. T.C. Chung and D. Rhubright, accepted in *J. of Polymer Science*.

SUPPORTED LEWIS ACID CATALYSTS

BASED ON POLYOLEFIN THERMOPLASTICS

T.C. Chung* and A. Kumar

Department of Materials Science and Engineering
The Pennsylvania State University
University Park, PA 16802

and

F.Chen and J. Stanat

Exxon Chemical Company
1900 East Linden Ave.
Linden, NJ 07036

INTRODUCTION

Immobilization of catalyst presents several important advantages in chemical production processes. These include waste reduction, lower cost and simpler purification. All of them also save energy and environment. These considerations become even more important when the reactions require a large quantity of catalyst, such as in the oligomerization of olefins. Figure 1 illustrates the reaction process. The catalyst involves in the oligomerization reaction and stays on the surface of the support by a stable chemical bond. Therefore, the product is a simple organic mixture, polyolefin and unreacted olefin and solvent, both can be easily recovered by distillation.

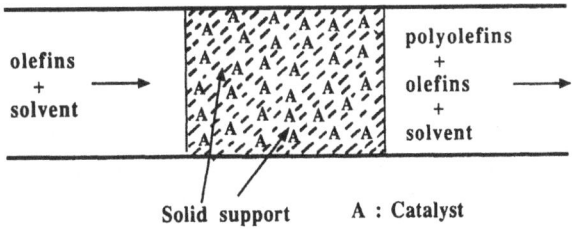

Figure 1. Pictorial scheme in fixed-bed process.

*Author to whom all correspondence should be addressed.

New Advances in Polyolefins, Edited by
T.C. Chung, Plenum Press, New York, 1993

Enormous research has been done on supported catalysts prepared by dispersing metal salts on inorganic substrates[1,2], such as magnesium oxide, alumina, silica, zeolite, followed by calcination/reduction[3]. In the fine powder form, these highly dispersed particles allow effective metal utilization. However, the supported catalysts are usually hydrolyzed and become part of the waste stream after a reaction cycle. In many cases, the control of the reactive sites is very difficult.

There are some examples of using polymers[4,5] as the solid substrates for both heterogeneous and homogeneous catalysts. The most common one is crosslinked polystyrene[6,7]. In homogeneous catalyst systems, the utilization of the insoluble polymer substrate makes it possible to recover the soluble catalyst from the reaction mixture. However, the activity of polymer-bonded catalyst is always relatively low due to diffusional limitation[8]. Some catalysts may not exist on the surface of substrate. In addition, the activity of most polymer-bonded catalysts decreases after a few reaction cycles. The reasons for the deactivation may be related to the insufficient stability of the polymer substrate and the slow diffusion of the catalyst into the polymer matrix. It is clear that there is a scientific challenge to develop a new type of polymer substrate which not only offers catalysts recoverability but also maintains long term stability and activity.

The use of a polyolefin as the substrate is very rare. This may be related to the difficulties in the preparation of functionalized polyolefins with appropriate and sufficient number of anchor sites to immobilize soluble catalysts. Ethylene oligomer, with a molecular weight of 1,000 g/mole, has been shown as an useful substrate for the preparation of polymeric reagents and catalysts.[9] In this case, only one active site exists in each polymer chain, the reaction usually takes place at high temperature and in a homogeneous phase. Upon cooling or addition of non-solvent, the catalysts bonded to ethylene oligomer become insoluble and can be recovered by filtration.

Polyisobutylene (PIB), both high and low molecular weight, is an important commercial polymer. The polymer is usually prepared by cationic polymerization of isobutylene[10] using Lewis acids, such as $AlCl_3$, $EtAlCl_2$, Et_2AlCl, BF_3 etc. as catalyst. The molecular weight of the polymer can be controlled by temperature,[11] solvent and catalyst concentration. The high molecular weight polymer (> 50,000 g/mole) requires very low polymerization temperature, such as - 95 °C. To obtain low molecular weight PIB (~ 2500-500 g/mole), the polymerization usually takes place at 0 °C. Chain transfer, involving β-proton elimination, is known to be a termination step which results in unsaturation. In other words, each PIB has one double bond. For preparing low molecular weight PIB, quite high concentrations of these catalysts have to be used, sometimes as high as 1%. It is certainly of great interest to study reusable Lewis acid catalysts, specially BF_3 which causes environmental concerns. However, BF_3 results in a desirable molecular structure[12] of PIB, containing a high percentage of terminal double bonds. To our knowledge, no example has been demonstrated on the immobilization of Lewis acids for carbocationic polymerization. Gates and coworkers[13,14] have reported using $AlCl_3$ for generating superacids on crosslinked polystyrene surface; however, the catalyst becomes deactivated very quickly by unknown pathways.

RESULTS and DISCUSSION

In this paper a new type of polymer substrate, functionalized polyolefin, is discussed. The original idea to use semicrystalline high molecular weight polyolefin, such as isotactic polypropylene or isotactic poly(1-butene), as the substrate was based on several considerations, mainly (1) the chemical stability of polyolefin which allows the application of the supported catalyst under more severe reaction conditions, (2) the crystallinity of polyolefin which not only offers good mechanical strength to maintain catalyst integrity under agitation, but also results in high and stable surface area of catalysts. The crystallinity of polyolefin may restrict heteroatom groups (active species) on the surfaces of crystalline phases.

A new functionalization chemistry, using a borane comonomer[15-17] in the polyolefins polymerization by Ziegler-Natta catalyst, offers desirable functionalized polyolefins. Two polymers, isotactic polypropylene with 5 mole % of hydroxy groups (PP-OH)[18] and isotactic poly(1-butene) with 12 mole % of hydroxy groups (PB-OH)[19], were used as substrates. Both functionalized polyolefins are semicrystalline with the polyolefin segments crystallize and highly branched segments with functional groups form amorphous phases on the surface of the crystalline phase. The functional groups are connected to the polymer backbone with flexible methylene units, which offer good mobility of the anchor sites for catalysts.

Immobilization of Lewis Acids on Semicrystalline Polyolefins

The hydroxylated polyolefins, polypropylene or poly(1-butene), were reacted with Lewis acids, such as $EtAlCl_2$, Et_2AlCl and BF_3. As shown in Equation 1,

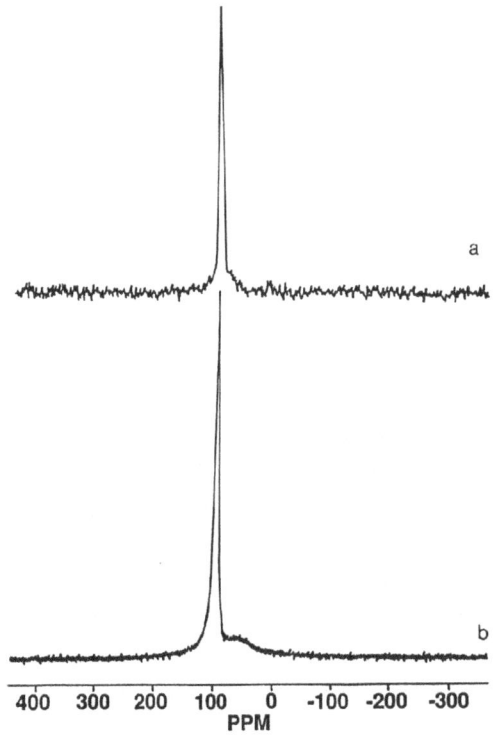

Equation 1

P is the partially crystalline polyolefins and M is a B or Al atom. The ligands (a, b, c) can be either alkyl or halogen groups and x, y can be a, b or c.

Figure 2. (a) Solid State ^{27}Al NMR spectrum of the catalyst A and (b) Solution ^{27}Al NMR spectrum of C_5-OAlCl$_2$.

The reaction was usually carried out at room temperature by stirring the hydroxylated polyolefin solid with excess Lewis acid solution for a few hours. The unreacted reagent is removed by washing the resulting supported catalyst with pure solvent several times. In general, the condensation reactions are quite complete, despite the shape and size of the polyolefin particles. For example, in the reaction between PB-OH and $EtAlCl_2$ elemental analysis shows that the concentration of Al species in the resulting supported catalyst (A) is very close to that of the hydroxy group in the original hydroxylated poly(1-butene). Most of the hydroxy groups must be located in the amorphous phases with good mobility and can be easily reached by $EtAlCl_2$ reagent.

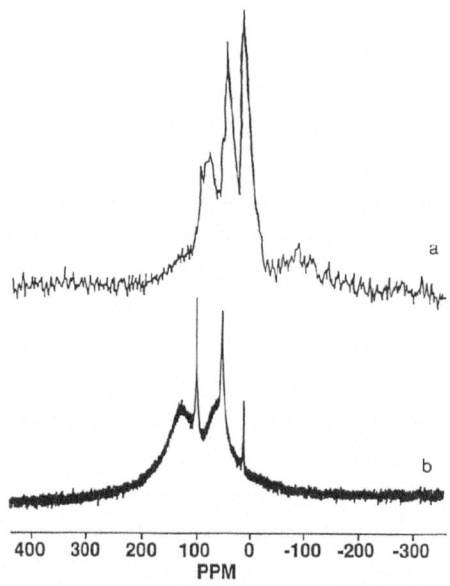

Figure 3. (a) Solid State ^{27}Al NMR spectrum of the catalyst B and (b) Solution ^{27}Al NMR spectrum of the adducts between 1-pentanol and Et_2AlCl.

Solid state ^{27}Al and ^{11}B NMR measurements were used to analyze the catalytic species in the supported catalyst. The spectra of three supported catalysts were compared with their corresponding soluble ones. Figure 2 (a) shows the ^{27}Al spectrum of catalyst (A), prepared by reacting hydroxylated polypropylene with $EtAlCl_2$ at room temperature. Only a single peak at 89 ppm, corresponding to $-OAlCl_2$,[20] was observed. There is no peak at 170 ppm, corresponding to $EtAlCl_2$. It is quite unexpected to discover such a selective reaction, the alkyl-aluminum bond is much more reactive than the aluminum-halide bond in the reaction with alcohol. In fact, the same chemical reaction was also observed in a reference sample, using 1-pentanol instead of the hydroxylated polypropylene in a reaction with a stoichiometric amount of $EtAlCl_2$. Figure 2(b) is the solution ^{27}Al NMR spectrum of the resulting $C_5-OAlCl_2$, which indicates the same chemical shift, corresponding to a single $-OAlCl_2$ species. The elemental analysis shown in Table I also implies the same result, with the similar theoretical and experimental mole ratio of Al/Cl in the $C_5-OAlCl_2$ compound.

In the case of Et$_2$AlCl, there are two alkyl groups which are both reactive to the hydroxy group. We expected that the resulting supported catalyst (B) would be a mixture. Figure 3 compares the solid state ^{27}Al NMR spectrum of the supported catalyst (B) to the solution ^{27}Al NMR spectrum of the corresponding small molecule formed by reacting 1-pentanol with the stoichiometric amount of Et$_2$AlCl. Both show similar results, with three main peaks at 93, 37 and 4 ppm, corresponding to -OAlEtCl , -O)$_2$AlEt and -O)$_2$AlCl respectively[23]. Again, no peak at 170 ppm corresponding to Et$_2$AlCl is observed. It is interesting to note that similar reaction mixtures were obtained for two completely different reaction conditions, homogeneous and heterogenous phases. The results indicate that the

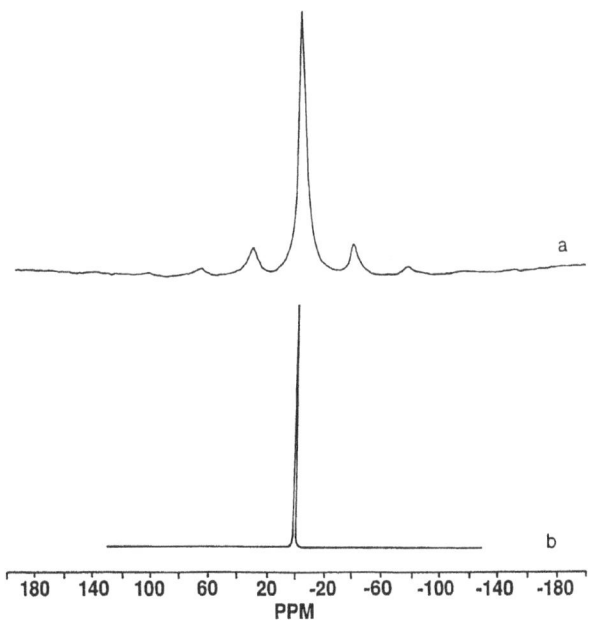

Figure 4. (a) Solid State ^{11}B NMR spectrum of the catalyst C and (b) Solution ^{11}B NMR spectrum of C$_5$-OBF$_2$.

hydroxy groups in solid poly(1-butene) are readily available during the reaction. Catalyst (C) is a supported BF$_3$ catalyst. The reaction between BF$_3$ and hydroxylated polymers (PB-OH) was done in two ways. The direct reaction between BF$_3$ and hydroxy group is very slow, and possibly forms BF$_3$/OH complexes. The more effective condensation reaction was carried out by using alkoxide groups. The metallization reaction of hydroxy groups was done by simple mixing polymer particles with n-butyl lithium solution. After washing out the excess n-butyl lithium, the polymer particles were mixed with a BF$_3$/CH$_2$Cl$_2$ solution. A similar procedure was applied to a control experiment using 1-pentanol. Figure 4 compares the ^{11}B NMR spectra of the two resulting catalysts.

Both spectra are almost identical with a peak at 0 ppm, corresponding to -OBF$_2$ group[21]. This result was also confirmed by elemental analysis, which shows a mole ratio of 1: 2 between B and F elements in the PB-OBF$_2$ sample.

Polyisobutylene Prepared by Polyolefin Supported Catalysts

The polymer supported catalyst (II) was used as the Lewis acid catalyst in the carbocationic polymerization of isobutylene as shown in Equation 2.

$$x \quad CH_2=C{\overset{CH_3}{\underset{CH_3}{}}} \quad \xrightarrow{\quad \text{(P)}-O\cdot M{\overset{X}{\underset{Y}{}}} \quad \text{(II)} \quad} \quad -(CH_2-C{\overset{CH_3}{\underset{CH_3}{}}})_{\overline{x}}$$

Equation 2

After the reaction, polyisobutylene (PIB) was obtained by filtering out the supported catalyst, then removing solvent and unreacted monomers under vacuum. The recovered catalyst was mixed with another isobutylene/hexane solution, followed by the same separation and recovery processes. This reaction cycle was repeated a number of times. Table I summarizes the results of PIB obtained by using the fine powder form (particle size < 1 mm) of catalyst (A), polypropylene supported -OAlCl$_2$ catalyst.

Typically, the monomer (isobutylene) to catalyst (-OAlCl$_2$) ratio was about 500 to 1. In most reaction cycles, the quantitative conversion from monomer to polymer was completed within 15 minutes. The same catalyst activity was maintained for several reaction cycles. This unusually high and stable catalyst activity in the polyolefin-supported system must be due to a

Table I. A Summary of PIB Prepared by PP-O-AlCl$_2$ (powder) in hexane

Run #	Temp (^{0}C)	Time (Min)	Mn (g/mole)	PDI	yield (%)
1	25	90	1,050	2.0	100
2	25	60	1,150	1.6	100
3	25	20	1,150	1.8	100
4	25	15	1,140	1.6	100
8	25	15	1,180	1.5	100
10	0	15	4,540	2.6	100
*	25	15	1,180	2.3	100
*	0	15	5,450	2.6	100

* C$_5$-O-AlCl$_2$ catalyst

Table II. A Summary of PIB Prepared by PP-O-AlCl2 (Chunk) in hexane

Run #	Temp (^{0}C)	Time (hr.)	Mn (g/mole)	PDI	yield (%)
1	25	3	1,370	3.0	100
2	25	2	1,660	2.6	100
3	25	1	1,230	2.4	65
4	25	3	1,110	2.6	100
8	25	2	1,400	2.4	92
14	25	3	1,320	2.6	100
15	0	5	5,450	2.6	76

high and stable surface area, which could be a consequence of small particle size, the crystallinity of polyolefin and the flexibility of the side chain between the catalyst and polymer backbone.

Table II shows results of PIB prepared by catalyst (A), with the same overall catalyst concentration. However, catalyst (A) was a big chunk, the particle size > 5 mm, instead of a fine powder form. The experimental results obtained over many consecutive reaction cycles are shown in Table II. In this case, the yield of PIB is very dependent on the reaction time. It requires about three hours to obtain complete conversion. This slow carbocationic polymerization of isobutylene is obviously related to the availability of catalyst. The big particle of PP-O-AlCl$_2$ greatly reduces the surface area of the catalyst, most of the active sites being buried inside the polymer matrix. The carbocationic polymerization took place in an extremely low catalyst concentration condition and most of isobutylene monomers had to travel some distance to the amorphous phases inside the polymer particle for the reaction to occur. Despite the difference in the reaction rate with the particle size, the catalyst can be recovered and reused in the subsequent reaction cycles. Elemental analysis and ^{27}Al NMR results show no significant change in the aluminum species after more than 10 reaction cycles.

In the cases of catalysts (B) and (C), the same recycle and reuse of the catalysts were observed in isobutylene polymerization. However, the catalyst activity and the resulting PIB structures, in terms of molecular weight and terminal unsaturation, were very different. As shown in Table III, catalyst (B) is much less active, requiring a long reaction time for completing the polymerization. Some of the aluminium species, -O)$_2$AlEt (five coordinations) and -O)$_2$AlCl (six coordinations) in catalyst B may contribute nothing to the polymerization. The major active sites could only be the -OAlEtCl species, which has a relatively low concentration.

On the other hand, the catalyst (C), PB-O-BF$_2$, is very active and produced relatively low molecular weight PIB, shown in Table IV, with terminal unsaturation. The details in molecular structure will be discussed in the following sections.

Table III. A Summary of PIB Prepared by Catalyst B (power)

Run #	Temp (^{0}C)	Time (hr.)	Mn (g/mole)	PD	Yield (%)
1	- 10	2	9,500	2.7	35
2	0	4	4,050	3.9	65
3	0	6	4,700	3.2	80
4	25	4	2,100	3.7	70
5	25	7	2,040	3.7	100
6	25	6	1,750	3.8	90
7	25	8	1,850	3.6	100

Table IV. A Summary of PIB Prepared by PB-O-BF$_2$ (Powder) in hexane

Run #	Temp (^{0}C)	Time (Min)	Mw (g/mole)	PDI	yield (%)
1	25	15	400	1.1	100
2	25	15	450	1.2	100
3	25	15	450	1.2	100
4	25	15	420	1.4	100
5	0	15	580	1.2	100
8	0	15	640	1.5	100
*	25	5	500	1.9	100
*	0	5	1080	2.0	100

(* C$_5$-O-BF$_2$ Soluble Catalyst)

Molecular Structure

In general, the molecular weight of PIB prepared by polyolefin-supported aluminium catalysts, A and B, is higher than those prepared by the conventional ones,[10] such as $AlCl_3$, $EtAlCl_2$ and Et_2AlCl. For examples, at 25 ^{0}C the molecular weight of PIB obtained using catalyst A was about 1100 g/mole and 2000 g/mole using catalyst B in hexane solution, compared to 500 g/mole in the case of $AlCl_3$ catalyst. The relatively higher molecular weight may be due to a slow chain tranfer reaction, because the carbocation in the supported catalyst is relatively stable. The alkoxide ligand donates π-electron density to the aluminum active site and stabilizes the propogating center. This effect can be very significant for the $-O)_2AlEt$ species in catalyst B. In fact, a high molecular weight peak (> 4000 g/mole) in the GPC curve was observed in the PIB sample prepared by catalyst B.

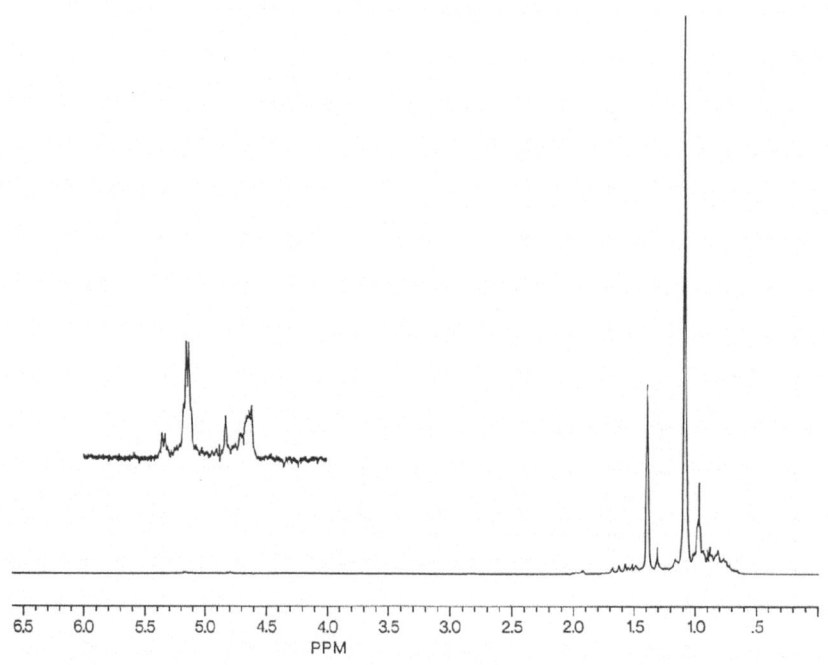

Figure 5. ^{1}H NMR spectrum of PIB prepared by catalyst A (PP-O-AlCl$_2$) at room temperature.

The molecular weight of PIB increased to 4,000 g/mol when the reaction temperature were decreased to 0° C, using catalyst A in hexane solution. The molecular weight-temperature relationship is expected as the general phenomenon[11] in carbocationic polymerization. It is interesting to note that a molecular weight (Mw=120,000 and Mn= 49,100 g/mol) of polyisobutylene can be obtained at relatively high temperature (-45 ^{0}C). In addition, the molecular weight of polyisobutylene is also dependent on the polarity of solvent, as in the conventional soluble cases[11]. All effects can be explained by the activity and stability of the active sites, which are dependent on ligand, temperature and solvent.

Figure 5 shows the ^{1}H NMR spectrum of PIB prepared by catalyst A (PP-O-AlCl$_2$) at room temperature. The overall spectrum is very similar to those of PIB prepared by soluble Al catalysts, such as $AlCl_3$, $EtAlCl_2$, Et_2AlCl, and the controlling C_5-O-AlCl$_2$ catalyst. Two major peaks, at 0.95 and 1.09 ppm, are due to CH_3 and CH_2 protons in the PIB polymer. There are some weak peaks located in the olefinic region, between 4.5 and 6.0 ppm.

This unsaturated double bond in the polymer chain is evidence that a proton chain transfer reaction (β-proton elimination) occurs in the polymerization process. There are two quartets at 5.4 and 5.2 ppm and two singlets at 4.9 and 4.6 ppm. The singlets at 4.9 and 4.6 ppm are indicative of two types of nonequivalent vinylidene hydrogens[22] which may be located at the end of polymer chain. The quartets at 5.4 and 5.2 ppm are the olefinic hydrogens coupled to the methyl group, which are due to the internal double bonds[23]. The ^{1}H NMR peak assignments are summarized in Table V.

Table V. Olefin Structures from ^{1}H NMR Shifts

Structure	Observed [1]H Chemical Shifts
C C ‖ ‖ -C-C-C-C=C ‖ C	4.88, 4.66 singlet
C C ‖ ‖ -C-C-C=C-C ‖ C	5.15 singlet
C C ‖ ‖ -C-C-C-C-C-C-C ‖ ‖ C C C	5.18 quartet
C C ‖ ‖ -C-C-C-C-C-C-C ‖ ‖ C C C	5.38 quartet

A significantly high amount of internal double bonds with various structures are present, which indicates that a carbocationic isomerization is taking place by the Lewis acid catalyst after the polymerization reaction. This olefin isomerization[22] is basically very similar to those in the soluble Al catalyst systems, such as $AlCl_3$, $EtAlCl_2$, Et_2AlCl.

A very different ^{1}H NMR spectra of PIB was observed by using supported catalyst C (PB-O-BF_2). As shown in Figure 6, the chemical shifts in the double bond region consist of two major singlets at 4.9 and 4.6 ppm, corresponding to terminal double bond, and a small peak at 5.15 ppm, corresponding to an internal double bond.

Comparing the integrated intensities between olefinic peaks, shown in Figure 6, the PIB prepared at 0°C contains more than 85% of terminal double bonds. The reason for such a high percentage of terminal double bonds in PB-O-BF_2 polymerization is not clear. The proton transfer reaction (β-proton elimination) is the termination step, as shown in Equation 3, which can form both terminal and internal double bonds.

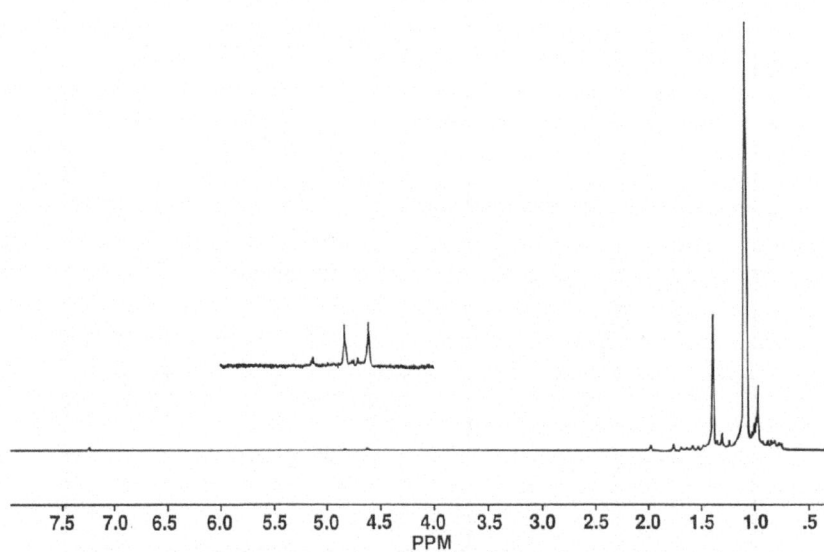

$$\text{wwwPIBwww} CH_2\text{-}\underset{\underset{CH_3}{|}}{\overset{\overset{CH_3}{|}}{C}}{}^{+} \ A^{-}$$

$\downarrow$ proton chain transfer

$$\text{wwwPIBwww} CH_2\text{-}\underset{\underset{}{|}}{\overset{\overset{CH_3}{|}}{C}}{=}CH_2 \quad + \quad \text{wwwPIBwww} CH{=}\underset{\underset{}{|}}{\overset{\overset{CH_3}{|}}{C}}\text{-}CH_3$$

Equation 3

Figure 6. ^{1}H NMR spectrum of PIB prepared by catalyst C (PB-O-BF$_2$) at 0 ^{0}C.

The elimination of a proton from two terminal methyl groups is statistically favorable. However, the elimination of a proton from the last methylene unit forms a thermodynamically stable internal double bond. Theoretically, the maximum percentage of terminal double bonds in the final PIB can not be more than 75%. It is reasonable to speculate that some effect due to the substrate may play a role in controlling the termination reactions and avoiding any isomerization reactions. In the control experiment using C$_5$-O-BF$_2$ soluble catalyst under the same reaction condition, more than 30 mole% of internal double bonds were formed.

It is interesting to note that the PIB product obtained with a high concentration of terminal double bonds is a very desirable material, called "reactive PIB", which can be further functionalized under mild reaction conditions. In the commercial process, the terminal double bonds react directly with maleic anhydride, and the internal unsaturated PIB requires halogenation reaction before maleic anhydride reaction. The current commercial "reactive PIB"[12] is a low molecular weight (500 - 2000 g/mole) PIB with 60 - 70% external double bonds, which is prepared by BF_3 catalyst. Functionalized PIB's are being used in many applications, such as additives in lubricating oil.

EXPERIMENTAL DETAILS

Materials and Measurements

The following chemicals, 9-borobicyclononane (9-BBN), $Al(Et)_3$, $AlEtCl_2$, $Al(Et)_2Cl$ and BF_3 (Aldrich), and $TiCl_3AA$ (Stauffer), were used as received. HPLC grade toluene, hexane and THF were distilled from sodium anthracide. Isopropanol and 1,5-hexadiene were dried with CaH_2 and distilled under N_2. Propylene (Matheson) was passed through P_2O_5 and NaOH columns before drying with $Al(Et)_3$ at low temperature. Isobutylene (Matheson) was used without further purification. All the manipulations were carried out in an innert atmosphere glove box or on a Schlenck line.

The molecular weight of polyisobutylene was determined using Waters GPC. The columns used were of Phenomenex Phenogel of 10^4, 10^3, 500 and 100 A. The flow rate of 0.7 ml/min was used and mobile phase was THF. Narrow molecular weight polystyrene samples were used as standards. All the solution NMR were done on a Bruker AM 300 machine. In the ^{27}Al NMR studies, toluene was used as solvent with deuterated toluene as the lock solvent. For ^1H NMR studies, deuterated chloroform was used as solvent. MAS ^{27}Al NMR were done at the CSU NMR Center on a Bruker AM 600 NMR spectrometer (^{27}Al resonance frequency of 156.4 MHz and 14.5 KHz MAS speed). MAS ^{11}B NMR were done on a Chemagnetics CMX 300 NMR spectrometer (^{11}B resonance frequency of 95.4 MHz and 4 KHz MAS speed).

Implantation of Aluminum Compounds ($EtAlCl_2$ and Et_2AlCl)

Two hydroxylated polymers, polypropylene containing 5 mole % hydroxy groups and polybutene containing 12 mole % hydroxy groups were used in the preparation of supported catalysts. Both polymers are slightly swellable in toluene. The reactions with the aluminum reagents were carried out at room temperature under an inert atmosphere. Both fine powder and big chunk of polyolefin particles were treated in the same way. In a typical example, the hydroxylated polyolefin (150 mg) polymer, suspended in toluene (15 ml), was mixed with excess aluminum compounds (~10 mmole). After a reaction time of 3 hrs, polyolefin was filtered and washed with hexane repeatedly to remove remaining aluminum compounds. Based on an elemental analysis and ^{27}Al NMR studies, most of the hydroxy groups were reacted, without any unreacted aluminium compound remaining in the polymer.

Implantation of Boron Trifluoride

In the reaction with BF_3, polyolefin containing hydroxy groups (150 mg) was reacted with a saturated solution of dichloromethane (15 ml) with boron trifluoride for 12 hrs. The excess boron triluoride and dichloromethane was removed under vacuum. The most effective method involves a pretreatment of hydroxylated polyolefin (150 mg) with 0.1 ml (1 mmole) of n-BuLi (10 M) in 7 ml of toluene for 1 hours. Polyolefin was filtered and washed with hexane to remove excess lithium compounds. The traces of solvent from the polyolefin powder were removed by vacuum. To this polymer a saturated solution of dichloromethane with BF_3 was added. This mixture was stirred at room temperature for 3 hours. Dichloromethane and excess BF_3 were removed on vacuum line.

Synthesis of C_5-O-$AlCl_2$ and C_5-O-BF_2

For the control reactions, the soluble catalysts are prepared by mixing pentanol (0.5 ml, 4.6 mmole) in 5 ml toluene before reacting with 0.48 ml (4.6 mmole) $EtAlCl_2$ which was diluted with 5 ml toluene. The solution of $EtAlCl_2$ was cooled to -78°C and to this cooled solution pentanol solution was added dropwise. It was stirred at -78°C for 15 minutes and then warmed up to room temperature. By removing toluene under vacuum, the C_5-O-$AlCl_2$ compound was obtained. The C_5-O-BF_2 was prepared by mixing 0.5 ml (4.6 mmole) of pentanol with 5 ml of ether. After the solution was cooled to -78°C, 0.45 ml (4.5 mmol) of n-BuLi (10 M) in 5 ml ether was added slowly. The reaction mixture was warmed up to room temperature and stirred for 10 minutes. Ether was removed under vacuum. To 200 mg of pentoxylithium in 5 ml dichloromethane, 20 ml of saturated solution of dichloromethane with BF_3 was added at room temperature. This mixture was stirred for 15 minutes. Dichloromethane and residual BF_3 was removed under vacuum to obtain C_5-O-BF_2.

Polymerization of Isobutylene

The polymerization was carried out in a high vacuum apparatus which consists of two 100 ml flasks (A and B) and one stopcock (a) was used to separate the flasks. The other stopcock (b) was used to control the vacuum conditions and nitrogen flow. In the dry box, a supported Lewis acid catalyst, such as 100 mg of catalyst (A), was charged to the flask A, the valve (b) was then closed. The whole apparatus was moved to a vacuum line and was pumped to high vacuum before closing the valve (b). Isobutylene (4ml, 50 mmole) was condensed in flask B and dissolved in about 20 ml hexane, which was vacuum-distilled into flask B. Isobutylene solution was warmed up to the required temperature and transferred to the catalyst in flask A. After stirring the reaction mixture for the required time, PIB solution was separated from the supported catalyst by filtration in the dry box. PIB was obtained by evaporating the solvent under vacuum. The supported catalyst was then recharged to flask A and the entire process was repeated.

CONCLUSION

This paper describes a new class of Lewis acid catalysts which are chemically bonded to the functionalized polyolefins, such as hydroxylated polypropylene and poly(1-butene). The catalysts are active in the carbocationic polymerization of isobutylene and can be recovered and reused for many reaction cycles without significantly losing their activity. Due to the crystallinity and brush-like molecular structure of the polyolefins, the Lewis acid catalysts are located on the surfaces of crystalline phases with high surface area and mobility. In addition, the use of partially crystalline polyolefin as the substrate offers several advantages, mainly, (a) chemical stability of polyolefin under severe reaction conditions, (b) good processibility of polyolefin, which offers a convenient way to prepare various shapes and sizes of substrate with high mechanical strength, (c) high catalyst loading to the substrate by simple implantation.

ACKNOWLEDGEMENTS

The authors would like to thank the financial support from the Polymer Program of the National Science Foundation and Exxon Chemical Company.

REFERENCES

1. Imamoglu, Y., "Olefin Metathesis and Polymerization Catalysts." Kluwer Publishers, 1990.
2. Karol, F. J., Mann, W. L., Goeke, G. L., Wagner, B. E., and Maraschin, N. L., J. Polym. Sci., Polym. Chem. Ed.**16**, 771 (1978).
3. Tanaka, K., J. Chem. Soc., Chem Commun., 748 (1984).
4. Chauvin, Y., Commereuc, D., and Dawans, F., Prog. Polym. Sci. **5**, 95 (1977).

5. Ford, W. T., Polymeric Reagents and Catalysts; ACS Symposium Series 308, 1986.
6. Manecke, G., and Storck, W., Angew. Chem. Int. Ed. Engl. **17**, 657 (1978).
7. Magnotta, V. L.; Gates, B. C. J. Catal. **46**, 266 (1977).
8. Ekerdt, J. G., Charpter 4 in Reference 6.
9. Bergbreiter, D. E., Chen, L. B., and Chandran, R., Macromolecules **18**, 1055 (1985).
10. Kennedy, J. P., and Marechal, E., "Carbocationic Polymerization." John Wiley & Sons, 1982.
11. Kennedy, J. P., and Trivedi, P. D., Adv. in Polym. Sci. **28**, 113 (1978).
12. Samson, J. N., US. Pat. 4,605,808 (1986).
13. Magnotta, V. L., Gates, B. C., and Schuit, G. C. A., J. Chem. Soc. Chem. Commun., 342 (1976).
14. Dooley, K. M., and Gates, B. C., J. Polym. Sci. Polym.Chem. Ed. **22**, 2859 (1984).
15. Chung, T. C., Macromolecules **21**, 865 (1988).
16. Ramakrishnan, S., Berluche, E., and Chung, T. C., Macromolecules **23**, 378 (1990).
17. Chung, T. C., ChemTech **27**, 496 (1991).
18. Chung, T. C., and Rhubright, D., Macromolecules **24**, 970 (1991).
19. Chung, T. C., and Rhubright, D., Macromolecules (accepted).
20. Benn, R., and Rufinska, A., Angew. Chem. Int. Ed. Engl. **25**, 861 (1986).
21. Noth, H., and Wrackmeyer, B, "Nuclear Magnetic Resonance Spectroscopy of Boron Compounds." Springer-Verlag, 1978.
22. Puskas, I., Banas, E. M., and Nerheim, A. G., J. Polym. Sci.Sym. No. 56, 1976,
23. Kastrup, R. V., Chen, F. J., and Emert, J., unpublished results.

NOVEL AZO INITIATORS FOR POLYOLEFIN MODIFICATION AND GRAFTING

Hai-Qi Xie and Warren E. Baker

Department of Chemistry
Queen's University
Kingston, Ontario
K7L 3N6, Canada

INTRODUCTION

Modification of relatively inexpensive and widely available polyolefins through free radical induced melt grafting is now recognized as a viable method for developing new polymeric materials without major capital investment. Such modified polymers generally contain polar and/or reactive functional groups which impart increased usefulness in various applications such as reactive compatibilization of immiscible polymer blends and tie layer adhesion in multilayer polymer products. New polymer blends with improved properties can be obtained through reactive blending using polymers functionalized with complementary coreactive groups.[1]

Although the melt grafting of monomers containing acidic and/or electrophilic functionalities such as carboxylic acid and anhydride groups onto preformed polymers has been extensively investigated,[2,3] few reports can be found in the literature concerning the functionalization of molten polymers with monomers containing the complementary coreactive groups, i.e., basic and/or nucleophilic functionalities. Since amines are the most common organic bases and/or nucleophiles, efforts have been made recently in the melt grafting of amine-containing methacrylates onto polyethylenes with organic peroxide initiators.[4-7] However, oxidation of the amino groups by organic peroxides occurred at the elevated temperatures experienced in melt grafting processes (160 - 220°C). Direct evidence of this oxidation was the intense discoloration (from yellow to brown) of the grafted polymers,[5] indicating the formation of amine oxides,[8] the oxidation products of the amines. This oxidation reaction detracts from the principal merits of the melt grafting process, since strongly colored polymers are of reduced commercial value. In addition, the amine oxides have no basicity nor nucleophilicity, reducing the reactivity of grafted polymers in blending applications. Therefore, non-oxidative initiators must be used in the melt grafting of amine-containing methacrylates onto polyethylenes.

Azo initiators are potential candidates as they are non-oxidative. However, the conventional nitrile azo initiators such as 2,2'-azobisisobutyronitrile (AIBN) are of very

New Advances in Polyolefins, Edited by
T.C. Chung, Plenum Press, New York, 1993

101

low thermal stability[9] and not capable of abstracting hydrogen atoms from polyolefin backbones.[10] These factors prevent common azo initiators from being used in melt grafting processes where high thermal stability and hydrogen-abstracting capability are required. This is evidenced by the experiments conducted by Simmons and Baker on the melt grafting of DMAEMA onto LLDPE using AIBN which showed no grafting.[4]

Among the non-conventional azo initiators reported in the literature, α-acetoxyazo initiators[11] are highly thermal stable and could be useful for melt grafting processes. In addition, the α-acetoxyazo initiators were claimed to cause crosslinking of a number of polymers including polyethylene.[11] However, Simmons and Baker reported that using 2,2'-azobis(2-acetoxypropane) (LUAZO AP) in the melt grafting of DMAEMA onto LLDPE gave irreproducible results.[4] This observation was explained by the fact that LUAZO AP has too long a half-life at the melt grafting temperature. For example, the half-life of LUAZO AP at 160°C is estimated to be 3900 min. Hence, azo initiators with more suitable thermal stability should be used.

In this study, the usefulness of three azo initiators with more suitable half lives under melt grafting conditions is tested with polyethylene crosslinking and polypropylene degradation experiments (called "hydrogen abstraction experiments"). These were followed with trials to graft DMAEMA and tBAEMA onto LLDPE. Only one of the three azo initiators, 2-phenylazo-2,4-dimethyl-4-methoxylvaleronitrile (V-19), is found to abstract hydrogen atoms from polyolefin backbones and induce grafting within a reasonable temperature range. The discoloration of grafted polymers is also monitored. The results for the melt grafting and the discoloration with V-19 are compared with those obtained with organic peroxide initiators. Influence of experimental conditions (monomer and initiator feed concentrations and processing temperature and time) on the melt grafting of tBAEMA onto LLDPE using V-19 is also investigated.

EXPERIMENTAL

Materials

LLDPE supplied by ESSO Chemical Canada (ESCORENE LL5103) is an ethylene-butene copolymer with a melt flow index of 12 dg/min, a density of 0.923 g/cm, a M_w of 58,000, and an M_w/M_n of 4. Proton NMR analysis indicated a comonomer content of about 7 wt% butene. The homopolymer polypropylene provided by Shell Canada Chemical Company (GE6100) has a melt flow index of 1 dg/min, M_w of 650 000 and M_w/M_n of about 14.

The monomers, DMAEMA from Aldrich and tBAEMA from Pfaltz & Bauer are used without further purification.

The azo initiators, 2-phenylazo-2,4-dimethyl-4-methoxyl-valeronitrile (V-19), 2-(carbamoylazo)isobutyronitrile (V-30), and 1,1'-azobiscyclohexylnitrile (V-40) were supplied by Wako Chemicals USA. V-19 is a transparent yellow liquid with a boiling point of 149 - 150°C / 1mmHg, V-30, a pale yellow crystalline solid with a melting point of 76 - 78°C, and V-40, a white solid with a melting point of 113 - 115°C. The organic peroxide initiators, 2,5-dimethyl-2,5-di(t-butylperoxy)hexane (L101), 2,5-di-methyl-2,5-di(t-butyl-peroxy)hexyne-3 (L130), and 1,1-di(t-butylperoxy)-3,3,5-trimethyl-cyclohexane (L231), supplied by Atochem North America, are clear liquids. The decomposition half-life data of V-19, V-30, V-40, L101, L130, and L231 in the useful temperature range for melt grafting processes are listed in Table 1.[12-13]

Table 1 Half-life (min) data of initiators in the temperature range useful for melt grafting processes

ITIATOR	$t_{1/2}$ (min)		
	160°C	180°C	200°C
V-19[a]	25	5.8	1.5
V-30[a]	2	0.3	0.06
V-40[a]	0.1	0.02	0.003
L101[b]	2	0.7	0.1
L130[b]	20	3	0.5
L231[b]	0.6	0.1	0.02

[a]Ref.12; [b]Ref. 13

Grafting Procedure

The melt grafting of DMAEMA and tBAEMA onto LLDPE is carried out on a Haake Buchler Rheomix 600 mixer, a batch-type internal intensive mixer with a capacity of 50 cm^3. The mixer is interfaced with a Rheocord 40 microprocessor, which allows control of the processing variables such as processing time, temperature, and rotor speed, as well as continuous recording of torque and temperature with time.

Prior to the melt grafting, the desired amounts of monomer and initiator are added to a certain mass of LLDPE, and manually mixed until a homogeneous wet paste is obtained. This wet paste is fed into the preheated mixer and allowed to react for specified processing time. The reaction temperature and processing time are preset at desired values before feeding the wet paste into the mixer. The rotor speed is kept constant at 60 rpm for all grafting experiments.

Hydrogen Abstraction Experiments

The hydrogen abstraction experiments are performed by a similar procedure as in melt grafting, but without monomer. Caution is taken to obtain reproducible torque values for each experiment.

Polymer Purification and Analysis

After the melt grafting, the grafted polymers are dried in a vacuum oven at 80°C for two days to constant weight in order to remove the unreacted monomer. The dried samples are dissolved in refluxing toluene at a concentration of ca. 0.04 g/cm^3, and the grafted polyethylene is selectively precipitated from the solution by addition to 10 volumes of stirred methanol. The operation of dissolving in toluene and precipitating in methanol is repeated until no homopolymer DMAEMA or tBAEMA remains in the grafted polymers. The degree of grafting (DG, wt% on PE) is measured by proton-NMR spectroscopy in toluene-d_8 at 90°C on a Bruker 400 MHz NMR spectrometer. The relative error of DG is determined to be at the 5% level.

The weight losses during selective precipitation are assumed to be homopolymer poly(tBAEMA) dissolved in methanol and the weight percentage of homopolymer (w_h, wt% of starting PE) is calculated from Eq. (1):

$$w_h = 100(W_d - W_p)(1 + DG/100)/W_d \qquad (1)$$

where W_d and W_p represent the weights of dissolved and purified samples respectively.

The grafting efficiency (GE, %), defined as the percentage of total converted monomer that is grafted, is calculated from Eq. (2):

$$GE = 100DG/(DG + w_h) \qquad (2)$$

The monomer conversion (x, %) is calculated through a mass balance using the following equation:

$$x = 100(DG + w_h)/c_m \qquad (3)$$

where c_m (wt% based on PE) is the monomer concentration in the feed.

The purified polymers are pressed into thin films at 180°C for 45 s under 14 MPa pressure. The thickness of the films is controlled between 0.1 - 0.2 mm. FTIR spectra of the thin films are obtained on a Bomen MB-120 FTIR spectrometer.

The colorimetric measurements of the grafted polymers are taken on a MacBeth 1500 ColorEye colorimeter using illuminant D_{65} with the UV component filtered out and with the specular component of the reflected light excluded. Injection molded disk specimens of 38 mm in diameter are used in the test.

RESULTS AND DISCUSSION

Hydrogen Abstraction Experiments

It is generally accepted that the first step in the melt grafting via free radical mechanism includes the abstraction of hydrogen atoms from polymer backbones by primary free radicals generated from initiator decomposition to create polymer macroradicals, the sites of the grafting reactions. However, in the absence of any monomer molecules, radical recombination, disproportionation, and/or ß-scission may occur, resulting in changes in polymer molecular weight depending on the nature of the particular polymer involved. For polyethylenes, crosslinking is the dominating reaction,[14] while for polypropylenes, degradation predominates.[15] Since changes in polymer molecular weight have direct influence on the viscosity of the polymer melt,[16] the torque exerted by the drive system on the polymer melt will vary accordingly in order to maintain the rotor speed constant. Therefore, the crosslinking of polyethylenes and the degradation of polypropylenes may be observed from torque traces recorded by the Rheocord 40 microprocessor during the hydrogen-abstraction experiments. The crosslinking of polyethylenes will be indicated by a torque increase and the polypropylene degradation, by a torque decrease.

Fig. 1 compares the torque variations with time during the precessing of LL5103 alone and with 0.9 wt% of azo initiator V-30 at 160°C for 8 min. It can be seen that a slight torque decrease is observed by the addition of V-30 into LL5103. V-40 has been shown to have a similar effect on torque values during processing with the same LLDPE at 140°C where its half-life is 0.5 min. Thus, the free radicals produced from the decomposition of V-30 and V-40 do not appear capable of abstracting hydrogen atoms from LLDPE. The small torque decreases can be explained by the plasticizing effect of the small molecules of the added initiators and their decomposition products. However, Fig. 2 clearly demonstrated that the addition of 0.9 wt% of V-19 into LL5103 results in ca. 60% torque increase after processing at 190°C for 8 min. This is evidence

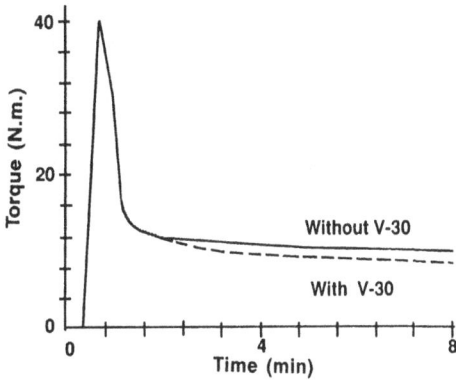

Fig. 1 Effect of added V-30 on torque during processing of LL5103
Conditions: 160°C, 100rpm, 0.9 wt% initiator

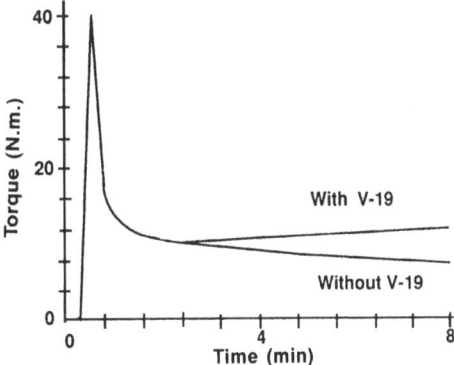

Fig. 2 Effect of added V-19 on torque during processing of LL5103
Conditions: 190°C, 100rpm, 0.9 wt% initiator

that the free radicals generated from V-19 decomposition can abstract hydrogen atoms from LLDPE resulting in polyethylene crosslinking. Comparable experiments using organic peroxide initiators L101, L130, and L231 also demonstrate torque increases indicating crosslinking of polyethylene. This is in agreement with other authors' findings that organic peroxide initiators are good hydrogens abstractors.[17]

It is well known that polypropylene is more sensitive to the attack of hydrogen-abstracting free radicals than polyethylene.[15] Thus one can confirm the observed difference in hydrogen-abstracting capability of V-19 compared to V-30 and V-40 by comparing torque decreases during the processing of molten polypropylene to which the particular initiator is added. The results obtained with V-19 and V-30 are presented in Figs. 3 and 4. It is clearly shown in Fig. 3 that the torque value for the processing of polypropylene GE6100 at 190°C for 8 min is reduced 81% from 9.0 to 1.7 N.m. by the addition of 0.5 wt% of V-19. However, Fig. 4 shows that the same amount of added V-30 causes only a 25% decrease in torque values for the processing of GE6100 at 180°C for 8 min. The minor torque reduction is primarily due to the plasticizing effect of the added small molecules as suggested in the polyethylene crosslinking experiments. V-40 has the same effect as V-30 on the processing of polypropylene. These polypropylene degradation experiments are further evidence that V-19 forms macroradicals on the polyolefin backbones while V-30 and V-40 cannot.

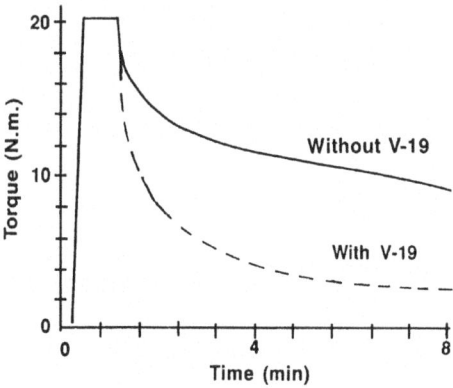

Fig. 3 Effect of added V-19 on torque during processing polypropylene GE6100 Conditions: 190°C, 100rpm, 0.9 wt% initiator

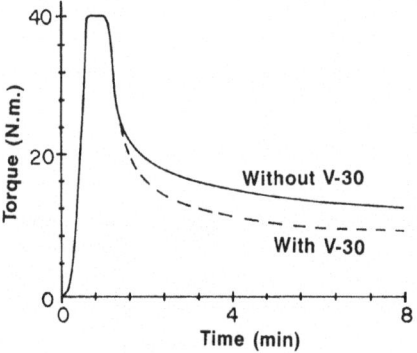

Fig. 4 Effect of added V-30 on torque during processing polypropylene GE6100. Conditions: 160°C, 100rpm, 0.9 wt% initiator

Characterization of Grafts

The azo initiators V-19, V-30, and V-40 are also used in the melt grafting of DMAEMA and tBAEMA onto LL5103. Figs. 5 and 6 present typical Proton-NMR and FTIR spectra of purified grafted LL5103 samples. As shown in Figs. 5-a and 6-a, the proton-NMR and FTIR spectra of LL5103 grafted with DMAEMA using V-30 have no new peaks relative to those of the virgin LLDPE. Similar proton-NMR and FTIR studies on the purified LL5103 grafted with DMAEMA and V-40, or tBAEMA and V-30 or V-40 do not show any evidence of grafting either. These results clearly demonstrate that the azo initiators V-30 and V-40 are not capable of initiating grafting reactions of DMAEMA or tBAEMA onto LLDPE. This observation is in agreement with the results obtained in the hydrogen abstraction experiments, in which the V-30 and V-40 are shown not to be capable of abstracting hydrogen atoms from polyolefins.

The observation that V-40 cannot abstract hydrogen atoms from polymer backbones and initiate grafting can be explained by the fact that its decomposition results in the formation of 1-cyanocyclohexyl radical. The latter is a tertiary radical not capable of abstracting hydrogen atoms from polymer backbones.[10] In the case of V-30, one of the two radicals generated from its decomposition is a tertiary butyl radical which is not a hydrogen-abstracting radical either. The other, the carbamoyl radical is unstable and decomposes easily yielding CO, CO_2, N_2, NH_3, etc.[10] Therefore, neither

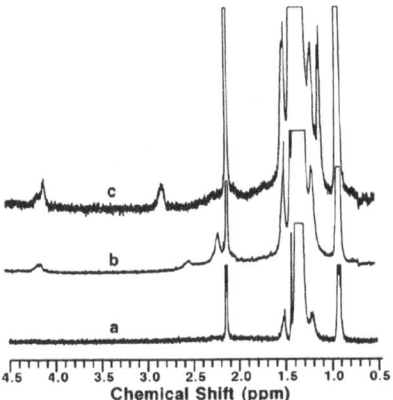

Fig. 5 Comparison of proton NMR spectra of virgin and purified grafted LLDPE: (a) grafted with DMAEMA using V-30; (b) grafted with DMAEMA using V-19; (c) grafted with tBAEMA using V-19. Shadowed peaks are not seen for virgin LL5103

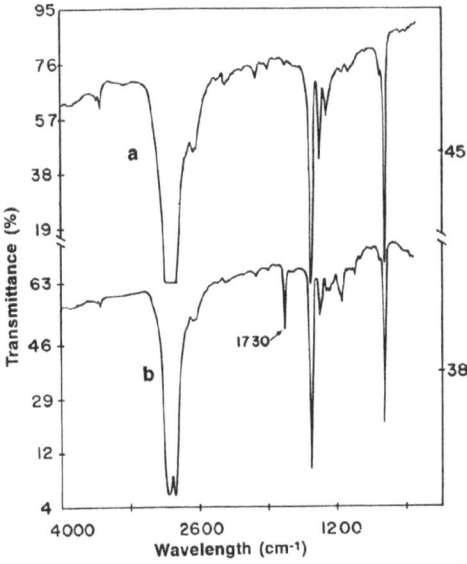

Fig. 6 FTIR spectra of purified grafted LL5103 films: (a) grafted with DMAEMA using V-30; (b) grafted with DMAEMA using V-19. Shadowed peaks are not seen for virgin LL5103

V-30 nor V-40 is capable of abstracting hydrogen atoms from polyolefin backbones and initiating grafting reactions.

However, the proton-NMR spectra of the purified grafted LL5103 with DMAEMA and tBAEMA using V-19 (Figs. 5-b and 5-c) obviously show new peaks not seen in the pure LLDPE. The broad peaks at 4.05 - 4.25 and 2.50 - 2.90 ppm indicate respectively the methylene protons adjacent to oxygen and nitrogen in the graft chains. In the case of DMAEMA grafts, the six methyl protons have resonance at 2.15 ppm

(Fig. 5-b). The nine t-butyl protons of tBAEMA graft resonate at 1.2 ppm (Fig. 5-c). In the FTIR spectrum of DMAEMA grafted polyethylene (Fig. 6-b), the ester carbonyl group is shown by its characteristic absorption centered at 1730 cm^{-1}. The C-O and C-N bendings are characterized by their typical absorption between 1000 to 1350 cm^{-1}. A similar FTIR spectrum is obtained for tBAEMA grafted LLDPE. Hence, grafting of DMAEMA and tBAEMA onto LLDPE is realized using the phenylazo initiator V-19. The results of these spectroscopic investigations are in agreement with the previously reported charaterization of the same grafts using peroxide initiators.[4-7]

Comparison of grafting with V-19, L101, L130 and L231

Since the azo initiators V-30 and V-40 cannot initiate the grafting, it is only possible to compare the grafting results obtained using initiators V-19, L101, L130, and L231.

Table 2 compares the DG values for grafting of DMAEMA and tBAEMA onto LLDPE using V-19 with those using L101, L130 and L231. It is observed that under similar conditions, the grafting with V-19 yields comparable DG's to those obtained

Table 2 DG values for melt grafting of DMAEMA and tBAEMA onto LLDPE using V-19, L101, L130, and L231[a]

Initiator	Monomer	DG (wt% on PE)		
		160°C	180°C	200°C
V-19	DMAEMA	1.91	0.95	0.62
	tBAEMA	2.25	1.06/1.12[b]	0.46
L10	DMAEMA	1.96	1.01	0.62
	tBAEMA	2.46	2.12	0.86
L130	DMAEMA	1.96		
	tBAEMA	2.51/2.71[b]		
L231	DMAEMA	3.20		
	tBAEMA	4.56		

[a] Conditions: 10 min, 40g LL5103, 10 g tBAEMA, and 0.25g V-19
[b] Results of two independent runs.

using L101 and L130. At 160°C, the peroxide initiator L231 demonstrates ca. 70% higher DG than V-19. However, L231 decomposes more than 40 times as fast as V-19 (Table 1). In fact, according to the half-life data in Table 1, L231 can reach essentially 100% decomposition at 160°C for 10 minutes, while V-19 decomposes only 24%. It also should be mentioned that one mole of L101, L130, and L231 can produce four moles of hydrogen-abstracting radicals,[13] while one mole of V-19 generates only one mole of the hydrogen-abstracting phenyl radical (cf. below), though the molecular weights of these initiators are about the same. This indicates the unusually high hydrogen-abstracting capability of the radical formed from V-19 decomposition.

Hydrogen-Abstracting Capability of Phenyl Radical

As mentioned in previous section, the first step of the free radical grafting involves the abstraction of hydrogen atoms from polyolefin backbones. It is traditionally believed that organic peroxide initiators produce, upon their decomposition, strong hydrogen-abstracting free radicals,[18] while free radicals generated from the decomposition of conventional azo initiators are poor hydrogen abstractors.[10] These conclusions can be verified by considering the nature and reactions of the free radicals concerned.

The hydrogen abstraction reactions involved in the functionalization of molten polyolefins are those from primary, secondary, and tertiary carbons. Table 3 presents the estimated equilibrium constants (K's)[19] at 180°C for such hydrogen abstraction reactions by free radical of interest. An equilibrium constant K larger than 1 indicates

Table 3 Estimated equilibrium constants for hydrogen abstraction reactions at 180°C

Radical	Carbon Type		
	Primary	Secondary	Tertiary
Ph•	6×10^6	2×10^8	1×10^{10}
RO •	1×10^3	7×10^4	6×10^6
CH_3•	8×10^2	2×10^4	2×10^6
RCH_2•	1	3×10^1	2×10^3
$(R)_2CH$ •	4×10^{-2}	1	9×10^1
$(R)_3C$ •	4×10^{-4}	1×10^{-2}	1

a thermodynamically favorable hydrogen abstraction reaction. Hence, peroxide and methyl radicals are strong hydrogen abstractors with K's greater than 800 for hydrogen abstraction reactions towards all the primary, secondary and tertiary hydrogens. Peroxide and methyl free radicals are generated from the thermal decomposition of organic peroxide initiators such as L101, L130, and L231 (Eq. 4) and the fragmentation of the t-butyl peroxide radical (Eq. 5).[20]

$$RO-OC(CH_3)_3 \longrightarrow RO\bullet + \bullet OC(CH_3)_3 \qquad (4)$$

$$\bullet OC(CH_3)_3 \longrightarrow \bullet CH_3 + CH_3COCH_3 \qquad (5)$$

Therefore, organic peroxide initiators are good hydrogen abstracting agents. However, the conventional azo initiators such as AIBN produces only tertiary carbon radicals which, according to K values in Table 3, have little hydrogen-abstracting capability. In addition, the half-lives of commonly used azo initiators are too short[9] to be useful for melt grafting.

According to the K values in Table 3, phenyl free radicals should have even higher hydrogen abstracting capability than peroxide and methyl free radicals. In fact, kinetic studies on the hydrogen abstraction reactions of phenyl free radical in both the gas phase[21] and solution[22] were reported in the 60's. The Arrhenius parameters are assembled in Table 4. It is shown that the activation energy E_a is more favorable for the hydrogen abstraction reactions with phenyl free radical than with methyl free radical, while the pre-exponential factors are in the same range. For example, comparison of the data obtained in the gas phase shows that the activation energy of the hydrogen abstraction reaction from cyclopropane (secondary hydrogens exclusively) by phenyl free radical is 18.4 KJ/mol less than with methyl free radical. Thus, from the point of view of both thermodynamics and kinetics, the hydrogen abstraction reaction with a phenyl radical is more favorable than with other radicals.

Table 4 Arrhenius parameters for hydrogen abstraction reactions with phenyl and methyl free radicals

Substrate	log A $(M^{-1}s^{-1})$		$E_a (KJmol^{-1})$	
	Ph•	Me•	Ph•	Me•
CH_4	8.9	8.8	46.5	59.9
cyclo-C_3H_6	8.4	9.0	35.6	54.0

The phenylazo initiator V-19 decomposes in a two-step process (Eq. 6),[23] yielding two

$$Ph-N{=}N-\underset{\underset{CN}{|}}{\overset{\overset{CH_3}{|}}{C}}-CH_2-\underset{\underset{OCH_3}{|}}{\overset{\overset{CH_3}{|}}{C}}-CH_3 \longrightarrow Ph-N{=}N\bullet + \bullet\underset{\underset{CN}{|}}{\overset{\overset{CH_3}{|}}{C}}-CH_2-\underset{\underset{OCH_3}{|}}{\overset{\overset{CH_3}{|}}{C}}-CH_3 \qquad (6)$$

$$Ph\bullet + N_2 \uparrow$$

types of radicals. The tertiary carbon radical produced in the first step cannot easily abstract hydrogen, but the phenyl radical can. Moreover, due to the stepwise nature of its decomposition, the tertiary carbon radicals may escape the solvent cage[10] before the formation of phenyl radicals. Thus, radical recombination within the solvent cage could be minimized (reduced caging effect) resulting in an enhanced initiator efficiency. This is particularly interesting for melt grafting, since polymer melts are generally highly viscous and the cage effects are likely to be very important. The considerations presented above suggest that the phenylazo initiator V-19 can be a more efficient initiator for polyolefin modification than traditionally used peroxide initiators.

Colorimetric Measurements

The goal of using an azo initiator instead of an organic peroxide is not limited to effecting grafting onto polyethylene but, more importantly, to reducing the discoloration of the grafted polymers. Thus, colorimetric measurements are performed on samples grafted with V-19 and peroxide initiators. Typical results (color indices) are listed in Table 5.

It is clearly shown that the light/dark and red/green indices of the grafted polymersare improved by using V-19 instead of organic peroxide initiators. In fact, the indices

Table 5 Comparison of color indices of DMAEMA grafted
LLDPE with V-19, L101, L130 and L231 at 160 C

	Color Index[a]		
Initiator	L	A	B
V-19	61	0	29
L101	56	5	30
L130	55	4	28
L231	54	4	30
None	64	-1	-7

[a] L: (100)Light/Dark(0)
A: (+)Red/Green(-)
B: (+)Blue/Yellow(-)

for polymers grafted with V-19 are very close to those of virgin LLDPE. However, addition of L101, L130 or L231 reduces the lightness and increases the redness of the polyethylene, which is consistent with the proposed formation of amine oxides, a red species. The blue/yellow index changes from -7 for unprocessed LLDPE to 29 ± 1 for polymers grafted with all the initiators studied.

Variables on Melt Grafting Using V-19

Table 6 summarizes the experimental conditions and results obtained (DG, w_h, GE and x) for the melt grafting of tBAEMA onto LLDPE using V-19. Thus, by varying the melt grafting conditions, DG's of up to 2.9 wt% are obtained with V-19. A detailed examination of the effects of various experimental conditions on the melt grafting are presented as follows.

Runs 1 - 7 are designed to study the effect of initiator concentration on the melt grafting at 180°C for 10 min. It is observed that both DG, w_h and x increase simultaneously with initiator concentration. Similar increases in DG, w_h and x were reported in the same melt grafting system using L130.[5] These increases can be attributed simply to the increasing number of free radicals being formed as more initiator is added. Nevertheless, GE has a maximum value at 0.63 wt% initiator concentration. This indicates that the maximum proportion of monomer converted to grafts occurs around the initiator concentration of 0.63 wt%. This initiator concentration is hence used in investigating the effects of most other experimental conditions on melt grafting.

The effect of monomer concentration on the melt grafting using 0.63 wt% V-19 at 180°C for 10 min is examined through Runs 8 - 13. It is observed that DG, x, and w_h increased continuously with monomer concentration. While these three parameters would have been expected to increase the GE also increses from 4.3 to 12% level for a monomer feed increase from 8 to 20 wt%. Further increase in monomer feed does not enhance GE significantly. This indicates that the homopolymerization is more predominant when a low tBAEMA concentration is used. This may be due to the alkyl radical R· generated from the first step of V-19 decomposition (Eq. 6), which easily initiates the homopolymerization of tBAEMA, but is not sufficiently energetic to abstract a hydrogen atom.[10]

Runs 14 - 20 are used to study the effect of reaction temperature on the melt

Table 6 Summary of melt grafting results of tBAEMA onto LLDPE using phenylazo initiator V-19

Run #	Temp. (°C)	Time (min)	c_m (wt%)	c_i (wt%)	DG (wt%)	w_h (wt%)	GE (%)	x (%)
1	180	10	25	0	0.4	0	0	2
2	180	10	25	0.13	0.3	3.2	8.6	14
3	180	10	25	0.25	0.5	4.4	10.2	20
4	180	10	25	0.43	0.8	6.2	11.4	28
5	180	10	25	0.63	1.1	7.7	12.5	35
6	180	10	25	1.00	1.3	-	-	-
7	180	10	25	1.25	1.4	12.6	10.0	56
8	180	10	8	0.63	0.1	2.2	4.3	29
9	180	10	15	0.63	0.4	4.3	8.5	31
10	180	10	20	0.63	0.6	4.5	11.8	26
11	180	10	25	0.63	1.1	7.7	12.5	35
12	180	10	30	0.63	1.3	9.4	12.1	36
13	180	10	39	0.63	2.1	12.5	14.4	37
14	140	10	25	0.63	1.3	8.5	13.3	39
15	160	10	25	0.63	1.7	11.3	13.1	52
16	180	10	25	0.63	1.1	7.7	12.5	35
17	190	10	25	0.63	0.7	4.6	13.2	21
18	200	10	25	0.63	0.6	3.4	12.8	16
19	210	10	25	0.63	0.4	2.6	13.3	12
20	220	10	25	0.63	0.4	2.6	13.3	12
21	160	2	25	1.25	1.4	1.9	42.4	13
22	160	3	25	1.25	1.6	3.4	28.0	20
23	160	4	25	1.25	1.9	6.0	24.1	32
24	160	5	25	1.25	2.1	6.7	23.9	35
25	160	6	25	1.25	2.3	7.9	22.7	41
26	160	7	25	1.25	2.3	7.3	24.0	38
27	160	10	25	1.25	2.8	7.8	26.4	42
28	160	15	25	1.25	2.6	8.1	24.3	43
29	160	20	25	1.25	2.9	8.3	25.9	45
30	160	30	25	1.25	2.6	8.6	23.2	45

grafting with tBAEMA and V-19 concentrations and processing time fixed at 25 wt%, 0.63 wt%, and 10 min respectively. It is shown that GE is not significantly altered with temperature (GE = 13.5 ± 1.5 %). While maxima in DG, w_h, and x are observed at 160°C. Either lowering or raising temperature from this optimal temperature will considerably reduce both DG and w_h. Since both initiator decomposition and chain propagation rates are generally reduced with lower temperature, the low values in DG, w_h, and x in the low temperature range can be readily explained by the formation of fewer and shorter graft and homopolymer chains. The fact that DG and w_h are reduced in the high temperature range could be due to the faster chain termination reactions at high temperature, resulting in short graft and homopolymer chain length. The short grafts and homopolymer molecules result in a lower value of x at high temperature. One can also notice that DG and w_h remains at ca. 0.5 and 3 wt% respectively at temperatures above 200°C. This suggests the approach to the ceiling temperature of tBAEMA. A similar effect of reaction temperature on melt grafting was reported previously. Song et. al.[5] noticed that for the melt grafting of DMAEMA onto LLDPE with L130, DG reached its maximum value at 140°C. This reflects the higher thermal stability of V-19 compared to L130.

The effect of processing time on the melt grafting with V-19 is investigated at 160°C (Runs 21 - 30) with a monomer and initiator concentrations of 25 and 1.25 wt% respectively. It is clearly demonstrated that DG, w_h, and consequently x, progressively increase with grafting time. The values of DG, w_h, and x reach plateaus at a grafting time of 10 min. This indicates more advanced initiator decomposition, monomer grafting and homopolymerization.

Data in Table 6 also show that the GE's for grafting times of 2 and 3 min are in the 30 to 40% range which are significantly higher than the GE values for longer processing times. Song et. al.[5] have also observed similar phenomenon, but with more pronounced higher GE values with short processing time for melt grafting of DMAEMA onto LLDPE using L130 at 160°C. It has been suggested through model studies[24] that poly(DMAEMA) has limited solubility in LLDPE and phase separation will occur at longer grafting times. At short reaction times and low monomer conversions, the melt grafting system remains homogeneous, but if separation occurs monomer may preferentially go to the poly(DMAEMA) phase and, furthermore, termination may be slower. The high GE value with short grafting time is particularly interesting for the continuous grafting processes in single and twin-screw extruders where short residence time is realistic. It has been reported[6] that the melt grafting of DMAEMA and tBAEMA onto LLDPE using L130 in a intermeshing twin-screw extruder showed much lower DG, GE, and x than in the batch-type mixer. While this was attributed to challenges in injecting and mixing in the twin screw extruders, it appears worthwhile to evaluate the V-19 for grafting in such a continuous device.

CONCLUSIONS

Polyethylene crosslinking, polypropylene degradation, and melt grafting experiments demonstrate unambiguously that the phenylazo initiator V-19 is an efficient initiator for polyolefins modification, while the azo initiators V-30 and V-40 are not. Colorimetric measurements show that the discoloration of grafted polymers using V-19 is significantly reduced compared to those functionalized with organic peroxide initiators. The optimal conditions for melt grafting of tBAEMA onto LLDPE are estimated as 20 wt% monomer feed, 0.63 wt% initiator concentration, 160°C reaction temperature, and 2 - 3 min processing time. Theoretical considerations reveal the high hydrogen-abstracting capability of phenyl free radicals generated from the decomposition of V-19. It is concluded that the phenylazo initiator may serve better in some polyolefin melt modification processes than organic peroxide initiators.

REFERENCES

1. N.C. Liu and W.E. Baker, Adv. Polym. Technol., 11:249 (1992).
2. B.C. Trivedi and B.M. Culbertson, "Maleic Anhydride", Plenum, New York (1982).
3. B.M. Culbertson, Encycl. Polym. Sci. Eng., Wiley, New York, 9:225 (1987).
4. A.Simmons and W.E.Baker, Polym. Eng. Sci., 29:1117 (1989).
5. Z.Song and W.E.Baker, Angew. Makromol. Chem., 181:1 (1990).
6. Z.Song and W.E.Baker, J. Appl. Polym. Sci., 41:1299 (1990).

7. Z.Song and W.E.Baker, POLYMER, 33:3266 (1991).

8. G.M. Loudon, "Organic Chemistry", Adison-Wiley, Menlo Park, California (1984).

9. J.C.Masson, "Polymer Handbook", 3rd Ed., Wiley, New York, II-1 (1989).

10 C.S.Sheppard, Encycl. Polym. Sci. Eng., Wiley, New York, 2:143, (1986).

11. K. Rauer, H. Hofmann, H. Schiller and C.S. Sheppard, US Patent 4,129,531 (assigned to Pennwalt Corporation, 1977).

12. Anon., "Azo Polymerization Initiators", Wako Chemicals USA (1983).

13. Anon., "Evaluation of Organic Peroxides from Half-life Data", Lucidol Pennwalt Co., Buffalo, New York (1983).

14. S.S. Labana, Encycl. Polym. Sci. Eng., Wiley, New York, 4: 385 (1986).

15. P. Blais, D.J. Carlsson, and D.M. Wiles, J. Polym. Sci. Part A-1, 10:1077 (1972).

16. Z. Tadmor and T.G. Gogos, "Principle of Polymer Processing", Wiley, New Yor (1979).

17. D.W. Yu, M. Xanthos, and C.G. Gogos, Adv. Polym. Technol., 10:163 (1990).

18. C.S.Sheppard, Encycl. Polym. Sci. Eng., Wiley, New York, 11:1, (1988).

19. T.L. Hill, "An Introduction to Statistical Thermodynamics", Dover Publications, New York (1986).

20. J.K. Kochi, "Free Radicals", Wiley, New York, N.Y. (1973). Page II-683.

21. (a) W. Fielding and H.O. Pritchard, J. Phys. Chem., 66:821 (1962); (b) F.J. Duncan and A.F. Trotman-Dickenson, J. Chem. Soc., 4672 (1962).

22. (a) R.F. Bridger and G.A. Russell, J. Am. Chem. Soc., 85:3754 (1963); (b) A.F. Trotman, "Advances in Free-Radical Chemistry", Ed. G.H. Williams, 1:1 (1965).

23. R. Kerber, O.Nuyken, and L. Weithmann, Chem. Ber., 108:1533 (1975).

24. J.B. Felmine, W.E. Baker, and K.E. Russell, Polym. Preprints, 32:181 (1991).

HOMOGENEOUS REACTION OF MALEIC ANHYDRIDE WITH LOW DENSITY
POLYETHYLENE IN SOLUTION IN AROMATIC HYDROCARBONS

Norman G. Gaylord,[1,2] Rajendra Mehta,[1] and
Achyut B. Deshpande[1]

[1]Gaylord Associates
New Providence, NJ 07974
[2]Charles A. Dana Research Institute for Scientists Emeriti
Drew University
Madison, NJ 07940

INTRODUCTION

    Radical catalysts have been used to promote the reaction between
maleic anhydride (MAH) and polyethylene under heterogeneous as well as
homogeneous conditions. The heterogeneous reactions have been carried
out with molten LDPE [1-4], in which MAH is insoluble and with LDPE
film suspended in acetic anhydride [5], while the homogeneous reactions
have been carried out with LDPE and HDPE in xylene solution [6-8], using
peroxidic catalysts at high concentrations and/or at temperatures where
the half-life is extremely short, conditions which are effective in
initiating the bulk homopolymerization of MAH [1,9].

    Although the heterogeneous reaction resulted in the crosslinking of
the LDPE, the homogeneous reaction in xylene solution gave uncrosslinked,
xylene-soluble MAH-containing PE [5,6]. Porejko et al. [6] concluded
that the reaction of MAH with PE in xylene at 110°C, in the presence of
benzoyl peroxide (BP) or azobisisobutyronitrile, proceeded through the
initiation of the homopolymerization of MAH by radicals generated on the
PE by the attack of radicals from the catalyst. However, Braun et al [8]
concluded that the reaction of MAH with LDPE in xylene at 139°C in the
presence of BP, proceeded without grafting or homopolymerization of MAH
and that the radical generated on the PE added one MAH molecule to yield
a PE-MAH· radical which terminated by coupling, disproportionation or
hydrogen abstraction.

    The failure to precipitate poly-MAH on the addition of chloroform to
the xylene-MAH-peroxide reaction mixture in the absence of LDPE, was considered by Braun et al [8] as evidence against the proposal of MAH homopolymerization from sites generated on the LDPE by radical attack.

    However, it has been shown [10] that MAH reacts with an alkylbenzene
in the presence of BP at 100°C to yield a saturated adduct wherein the
point of attachment between a succinic anhydride moiety and the aromatic
hdyrocarbon is the benzylic carbon atom. In some cases, products from the
reaction also include adducts containing more than one MAH unit [11]. This
indicates that MAH homopolymerization can occur to form short MAH sequences.
The failure of Braun et al. to recover poly-MAH by precipitation from xylene

solution by the addition of chloroform, results from the relatively low
molecular weight of the poly-MAH sequences, actually containing only one
MAH unit, and the solubilizing effect of the aromatic moiety.

The present investigation was undertaken to examine the solution
polymerization of MAH in xylene and to determine whether poly-MAH and/or
xylene-MAH adducts are formed in the presence of LDPE. The solution
polymerization in benzene, which is devoid of benzylic carbon atom, was
also examined.

RESULTS AND DISCUSSION

Reaction in Absence of Solvent

The reaction of MAH with molten LDPE in the absence of a solvent and
in the presence of a peroxide, at a temperature where the latter is under-
going rapid decomposition, proceeded in a heterogeneous environment due to
the insolubility of MAH in aliphatic hydrocarbons. Poly-MAH was not recovered from the reaction mixture and crosslinking of the LDPE accompanied
the appendage, i.e. "grafting" of individual units derived from MAH onto
the LDPE [9].

Reaction in Xylene

In absence of LDPE

The homogeneous reaction of MAH with benzoyl peroxide in xylene, in
the absence of LDPE, at 120°or 140°C yielded a mixture of poly-MAH and
xylene-MAH adduct. When DMF was present in the homogeneous solution, only
the xylene-MAH adduct was obtained, free of poly-MAH (Table 1).

Table 1. Homopolymerization of MAH in aromatic solvents

MAH, g	4.9	1.0	1.0	4.9	4.9
BP, wt-% on MAH	7.4	7.4	7.5	7.4	7.4
Xylene, ml	50	60	60	0	0
Benzene, ml	0	0	0	50	5
DMF, ml	0	0	5	0	0
Temperature, °C	140	120	120	80	80
Time, hr	6	5	5	6	6
Products					
Adduct, %	42.0	39.5	49.9	0	0
Poly-MAH, %	3.1	49.5	0	12.4	22.4
MAH content, wt-%					
Adduct	40.9				
Poly-MAH		36.2			

The adduct and the poly-MAH were hydrolyzed and then treated with an
ethereal solution of diazomethane. The methyl ester of the MAH oligomer
had a molecular weight (vpo) of 916, representing a DP of 6. The molecular
weight of the methyl ester of the adduct was 225 (calculated 232).

In presence of LDPE

The decomposition of BP in a homogeneous solution of MAH and LDPE at
elevated temperature resulted in the appendage of individual MAH units to
the LDPE. Although the LDPE-g-MAH, containing about 1-2 wt-% MAH, was
accompanied by the xylene-MAH adduct in the absence as well as in the
presence of DMF, no poly-MAH was recovered when LDPE was present in the

Table 2. LDPE-MAH reaction in xylene[1]

LDPE, g	5.0	5.0	8.0	8.0
MAH, g	1.0	1.0	4.0	4.0
BP, wt-% on MAH	7.5	7.5	7.4	7.4
Xylene, ml	60	60	200	200
DMF, g	0.0078	0	0	0.029
Temperature, $^{\circ}$C	120	120	140	140
Time, hr	5	5	6	6
Products				
Adduct, % on MAH	59.6	45.7	30.9	34.4
Poly-MAH, % on MAH	0	0	0	0
LDPE-g-MAH				
Xylene-soluble				
% of total	97.5	98.3	97.8	98.6
MAH content, wt-%	1.75	1.85	2.27	1.24

[1] LDPE-xylene mixture, in absence or presence of MAH, heated at reaction temperature until LDPE dissolved; BP, accompanied by MAH, if latter not present in original reaction mixture, added and heating continued for 5-6 hr

homogeneous reaction mixture, independent of the absence or presence of DMF (Table 2).

The homogeneous solution of reactants was obtained by immersing a flask containing xylene and LDPE in a constant temperature bath at the desired reaction temperature. When the polymer had dissolved, generally after 1-2 hr, a mixture of BP and MAH, if the latter was not present in the original reaction mixture, was added to the solution. When all of the reactants, including the BP and MAH, were charged at room temperature and the flask was then placed in the constant temperature bath, the reaction mixture remained heterogeneous until the polymer dissolved. Since the BP was decomposing and generating radicals while the temperature was rising, polymerization of the MAH resulted in the formation of poly-MAH in the heterogeneous mixture containing undissolved LDPE, until a homogeneous solution was obtained by solution of the LDPE.

In order to recover the reaction products, the reaction mixture was poured into acetone. No poly-MAH was precipitated on the addition of chloroform to the filtrate from the homogeneous reaction nor was poly-MAH recovered on evaporation of the xylene, acetone and chloroform from the filtrate. However, the xylene-MAH adduct was recovered after evaporation of the filtrate. The solid was purified by sublimation of unreacted MAH.

Reaction in Benzene

In absence of LDPE

The decomposition of BP or t-butylperoxypivalate at 80° and 100°C in a homogeneous solution of MAH in benzene, resulted in the formation of poly-MAH (Table 1). The yield decreased with dilution, i.e. as the solvent content increased. No benzene-MAH adduct, analogous to the xylene-MAH adduct, accompanied the poly-MAH which separated from the reaction mixture when the latter was permitted to stand at room temperature. This is due to the absence of a labile or benzylic hydrogen in benzene, to participate in adduct formation.

In presence of LDPE

The decomposition of BP in a homogeneous solution of LDPE in benzene at 100°C yielded LDPE which was essentially completely soluble in xylene, indi-

Table 3. LDPE-MAH reaction in benzene

LDPE, g	2.5	2.5	2.5	10.0
MAH, g	0	0.25	0.25	1.0
BP, wt-% on MAH	0	10	10	10
wt-% on PE	1.0	1.0	1.0	1.0
Benzene, ml	25	25	25	100
DMF, ml	0	0	0.002	0
Temperature, $^\circ$C	100	100	100	100
Time, hr	2	2	2	2
Products				
Poly-MAH, % on MAH	0	0	0	0
LDPE-g-MAH				
Xylene-soluble				
% of total	99.0	85.0	98.5	85.6
MAH content, wt-%		1.14	1.61	1.44
Xylene-insoluble				
% of total		14.0		13.6
MAH content, wt-%		4.71		2.12

cating the absence of crosslinking of the LDPE under these conditions
(Table 3). However, the presence of MAH resulted in the formation of
crosslinked LDPE-g-MAH as well as xylene-soluble polymer. No poly-MAH
was formed.

The products from the LDPE-MAH reaction in benzene in the presence of
a peroxide, were precipitated by pouring the reaction mixture into acetone.
After the filtrate separated from the precipitated LDPE-g-MAH was evaporated,
the residue was dissolved in acetone but no poly-MAH was precipitated on the
addition of chloroform. Monomeric, unreacted MAH, if present, was removed
from the products by sublimation.

The formation of xylene-insoluble LDPE-g-MAH in the presence of BP and
MAH, but not in the absence of MAH, is analogous to the formation of cross
linked polymer in the bulk reactions of LDPE-MAH-BP in the melt, i.e. the
crosslinking results from the presence of polymerizing MAH which involves
excited MAH. The presence of DMF in the reaction mixture containing benzene
prevents the crosslinking, as it does in the bulk reaction.

SUMMARY AND CONCLUSIONS

The homogeneous reaction of MAH with BP in xylene at 120 and 140°C
yielded a mixture of poly-MAH and a 1:1 xylene-MAH adduct. The presence of
DMF in the homogeneous mixture resulted in the formation of the adduct, free
of poly-MAH. The decomposition of BP in a homogeneous xylene solution of
MAH and LDPE resulted in the appendage of individual MAH units to the LDPE.
Although the LDPE-g-MAH containing about 1-2 wt-% MAH, was accompanied by the
xylene-MAH adduct, in the presence or absence of DMF, no poly-MAH was
recovered when LDPE was present in the xylene solution.

The reaction of MAH with BP in benzene at 100°C yielded poly-MAH. No
benzene-MAH adduct was recovered. The decomposition of BP in a homogeneous
solution of LDPE in benzene yielded xylene-soluble LDPE. The presence of
MAH in the LDPE-BP-benzene solution yielded xylene-soluble LDPE-g-MAH, accompanied by crosslinked polymer. The presence of DMF prevented the formation
of crosslinked polymer. No poly-MAH was recovered when LDPE was present.

Although poly-MAH is generated in the absence of LDPE, the presence of
the PE prevents the formation of poly-MAH. It has been proposed by Gaylord
et al. [4] that monomeric MAH is converted to an excited dimeric species

which attacks the PE and generates a radical thereon. The latter, as well as radicals from attack by catalyst radicals, adds an excited MAH dimer. In the absence of LDPE, the dimer undergoes homopolymerization. In the presence of LDPE, the dimer adds to the radical on the polymer.

Since MAH homopolymerization can and does occur in the absence of LDPE, the appendage of MAH to the polymer without the formation of poly-MAH in the xylene solution, suggests that the reaction of MAH with the dissolved LDPE is preferred over reaction with propagating MAH chains, if any. This supports the conclusion that the appended MAH is present as individual MAH or succinic anhydride units rather than as short oligomeric poly-MAH chains.

It has been proposed that individual MAH units result from depolymerization of appended poly-MAH chains, because the reaction, particularly reactions with molten PE, are conducted above the ceiling temperature of poly-MAH. However, single units are incorporated even when the LDPE-MAH reaction is conducted far below the ceiling temperature. Thus, the latter plays little or no role in the determination of the structure of LDPE-g-MAH.

REFERENCES

1. N.G. Gaylord, J. Macromol. Sci., Revs. Macromol. Chem., 13, 235 (1975).
2. N.G. Gaylord and S. Maiti, J. Polym. Sci., Polym. Lett. Ed., 11, 253 (1973).
3. N.G. Gaylord and J.Y. Koo, J. Polym. Sci., Polym. Lett. Ed., 19, 107 (1981).
4. N.G. Gaylord, M. Mehta, and V. Kumar, Proc. Org. Coat. Appl. Polym. Sci., 46, 87 (1982).
5. W. Gabara and S. Porejko, J. Polym. Sci., A-1, 5, 1539 (1967).
6. W. Gabara and S. Porejko, J. Polym. Sci., A-1, 5, 1547 (1967).
7. S. Porejko, W. Gabara, and J. Kulesza, J. Polym. Sci., A-1, 5, 1563 (1967).
8. D. Braun and U. Eisenlohr, Angew. Makromol. Chem., 55, 43 (1976).
9. N.G. Gaylord and M. Mehta, J. Polym. Sci., Polym. Lett. Ed., 20, 481 (1982).
10. W.G. Bickford, G.S. Fisher, F.G. Dollear, and C.E. Swift, J. Am. Oil Chem. Soc., 25, 251 (1948).
11. H. Shechter and H.C. Barker, J. Org. Chem., 21, 1473 (1956).

STRUCTURE, CRYSTALLIZATION AND MELTING
OF LINEAR, BRANCHED AND COPOLYMERIZED POLYETHYLENES
AS REVEALED BY FRACTIONATION METHODS AND DSC

Vincent B.F. Mathot

DSM Research
P.O. Box 18
6160 MD Geleen
The Netherlands

ABSTRACT

The influence of chain length and short chain branching on crystallization and melting behaviour, as studied by differential scanning calorimetry (DSC), is discussed for linear polyethylene (LPE), high density polyethylene (HDPE), low density polyethylene (LDPE), homogeneous ethylene-propylene (EP) and ethylene-1-octene (EO) copolymers and heterogeneous ethylene-1-butene and ethylene-1-octene copolymers such as linear low density polyethylene (LLDPE) and very low density polyethylene (VLDPE). Attention is paid to the clarification of molecular structure, in particular the distribution of short chain branching over the chains. Other subjects discussed are the use of various methods of separating polymer chains according to molecular dimension, for example direct extraction (DE) and analytical and preparative size exclusion chromatography (SEC), and the complementary separation according to degree of short chain branching, which is actually a separation according to ethylene sequence length and is based on the crystallizability/dissolvability of chains. In the context of the latter separation, the use of techniques such as analytical and preparative temperature rising elution fractionation (TREF) is discussed. Cross-fractionation, which should preferably comprise a first separation according to molecular dimension followed by a separation according to

New Advances in Polyolefins, Edited by
T.C. Chung, Plenum Press, New York, 1993

crystallizability/dissolvability, is shown to be an important tool in obtaining insight into the detailed structure of heterogeneous copolymers.

INTRODUCTION

In the past decade, long-established polyethylenes such as linear polyethylene, LPE, high-density polyethylene, HDPE, and low-density polyethylene, LDPE, have been improved ever further, while increasing attention has been paid to ultra-high molecular weight polyethylene, UHMWPE, in both its linear and its branched forms. At the same time, the range of commercial ethylene-based copolymers has been greatly extended. Linear low density polyethylene, LLDPE, was successfully (re)introduced as a heterogeneous copolymer, and this paved the way for the introduction of very (ultra) low density polyethylene, VLDPE (ULDPE). Besides the heterogeneous copolymers, homogeneous copolymers are also increasingly becoming a focus of interest. And long-established ethylene-propylene terpolymers such as EPDM, which can in principle be polymerized heterogeneously as well as homogeneously, are now being produced on a world scale.

In the case of copolymers, the properties of the ultimate product can be influenced drastically by judiciously varying the molecular structure within the polymer chain and between chains, in particular by varying the way in which the comonomer units are added to the growing chains, and by optimizing the processing step. The type of comonomer is also an important parameter. Therefore, it is very important to establish the causal relationships between molecular structure (as determined by the catalyst and the polymerization conditions used) via morphology (which, besides to molecular structure, is related to crystallization and processing) to the ultimate property mix and vice versa. In the past decade many attempts have been made to unravel the molecular structure of heterogeneous copolymers. This inevitably involved the development and application of — often laborious — fractionation techniques and methods of analysing the fractions. The molecular structure has a large influence on crystallization and melting behaviour, so it is not surprising that DSC has proved to be an important tool in the analysis of the fractions, alongside methods such as IR/NMR and [η]/SEC.

This paper will discuss this type of structural analysis for the various polyethylene types and the influence of structure on crystallization and melting behaviour[1-9].

CLASSIFICATION OF POLYETHYLENES

Figure 1 shows a classification of various important polyethylene types. On the left-hand side, the density at room temperature has been used as a classification criterion. It is a well-known fact that LPE can be prepared in almost completely crystalline form, with a density of about 1000 kg/m^3 at room temperature, whereas the other end of the density

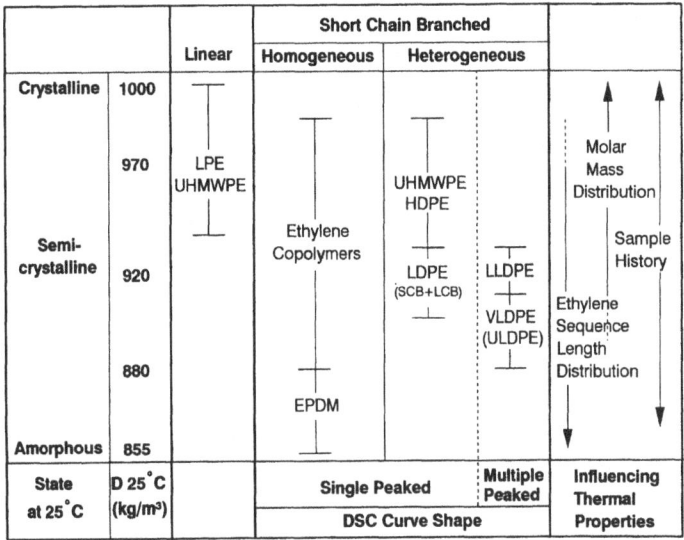

| | | Linear | Short Chain Branched | | |
			Homogeneous	Heterogeneous	
Crystalline	1000	LPE UHMWPE	Ethylene Copolymers	UHMWPE HDPE LDPE (SCB+LCB)	LLDPE VLDPE (ULDPE)	Molar Mass Distribution / Sample History / Ethylene Sequence Length Distribution
Semi-crystalline	970					
	920					
	880		EPDM			
Amorphous	855					
State at 25°C	D 25°C (kg/m³)		Single Peaked		Multiple Peaked	Influencing Thermal Properties
			DSC Curve Shape			

Figure 1. Classification of important polyethylene types according to overall density at room temperature, molecular structure, DSC-curve shape and parameters that influence the crystallization and melting behaviour.

range is formed by amorphous EPDM-rubbers with a density of about 855 kg/m³. Densities falling between these extremes are averages and represent semi-crystalline samples in which at least two phases (crystalline and amorphous) contribute to the average. These phases need not always be homogeneously distributed in the sample.

In LPE and linear UHMWPE, the main structural parameter influencing crystallization and melting is of course the molar mass distribution. In addition, the sample history is very important. The term "sample history" is used here in a very general sense and includes the polymerization process, crystallization during polymerization, working up and processing to obtain the final product, including the use of additives. An example of the great influence of both molar mass and sample history is of course the strong PE fibre based on UHMWPE. An important part of the sample history is the thermal history, which includes the temperature profile as a function of time. The figure shows in diagrammatic form in what range the above-mentioned parameters affect the density.

Short chain branched polyethylenes can be given any density between the two extremes mentioned by varying the amount of branching introduced in the ethylene backbone, for example by copolymerization. Moreover, at any degree of branching the density is further influenced not only by molar mass and sample history but also by the type of short chain branching (SCB), the distribution of branching in the chains and between chains, etc.

The short chain branching distribution (SCBD) in chains and between chains is actually the most important parameter for the crystallization and melting behaviour and thus partly determines the final property mix. No wonder that currently much attention is being paid to ways of influencing this distribution.

For the analysis of polymerized samples complex and laborious fractionation techniques are required. The analysis of the fractions obtained is moreover often difficult on account of the small amount of material in each fraction.

One of today's major problems is the determination of the SCBD or, in other words, the determination of the ethylene sequence length distribution (ESLD) in the chains. This is due to the fact that with ^{13}C-NMR only very short sequences (up to and including 5 methylenes besides the pooled methylenes in methylene sequences of length 6 and longer) can be determined quantitatively, while other techniques, for example those based on pyrolysis, are not (yet) capable of providing quantitative information about the longer sequences.

This is one of the reasons why DSC has become so popular as a characterization technique: the crystallization and melting behaviour of a polymer is influenced very strongly by the ESLD and this behaviour can be recorded relatively simply by DSC, while only a very small amount of material is required for the analysis. However, the problem of quantitatively translating this influence to molecular structure has not yet been solved for any of the polyethylene types mentioned above, although there have been many promising attempts. As an illustration, Figure 1 indicates the DSC curve shapes for the various polyethylenes.

The figure distinguishes between homogeneous and heterogeneous polyethylenes, because these differ considerably in terms of SCBD/ESLD. This distinction will be elucidated separately later; it coincides with very great differences in molecular structure and hence in crystallization and melting behaviour and properties. The figure shows that in spite of these differences the (average) density of the polymers concerned can be made the same; this illustrates, of course, that quantities such as density, degree of branching, heat of fusion, etc. are overall quantities, which provide only a superficial impression of the actual situation. More detailed information will have to be obtained from various distributions, such as molar mass distribution, sequence length distribution, crystallization temperature distribution, melting temperature distribution, etc.

Figure 2 shows, with reference to some linear and branched polyethylenes, the influence of (besides SCB[10,11]) molar mass[12,56] and cooling rate[13] on overall density as determined in a gradient column at room temperature.

The LPEs discussed here show a great influence of molar mass on density: variation of the viscosity-average molar mass, M_v, between about 3 kg/mol and 120 kg/mol results in a difference in density of about 30 kg/m^3, and it is a well-known fact that the density decreases further as the molar mass increases further.[56]

In addition, an increase in cooling rate, from slow (0.2 and 5 °C/min in a Mettler FP 52-oven) via 50 °C/min (LPE/HDPE/LLDPE) or 40 °C/min (LDPE) (average rates during crystallization in a press at a pressure of 180 kN/cm^2) to quenching (in a quench bath), results in a decrease in density (about 25 kg/m^3 for the LPEs, which decrease is

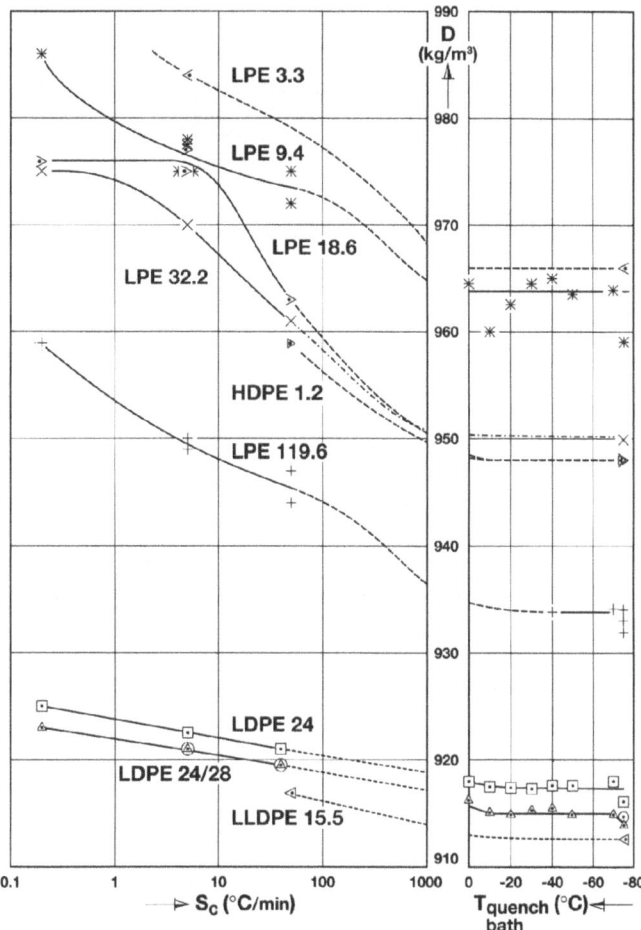

Figure 2. The influence on overall density at room temperature of molar mass and cooling rate for LPEs (numbers indicate M_v in kg/mol) and of cooling rate for an HDPE, some LDPEs and an LLDPE (numbers indicate CH_3/1000 C).

virtually independent of the molar masses considered here). The LDPEs and the LLDPE show a smaller influence of the cooling rate.

The samples (rods or films) were mounted in sample holders that were typically 1-3.5 mm thick. The temperature of the quench bath (quenching from 150 °C) was set with the aid of a cryostat with ethanol (temperatures down to -30 °C) or with ethanol with solid CO_2 (temperatures from -30 to -78 °C). In the procedure followed, the lower quench bath temperatures apparently did not result in faster cooling.

We shall now discuss the various distributions mentioned above for the various polyethylene types.

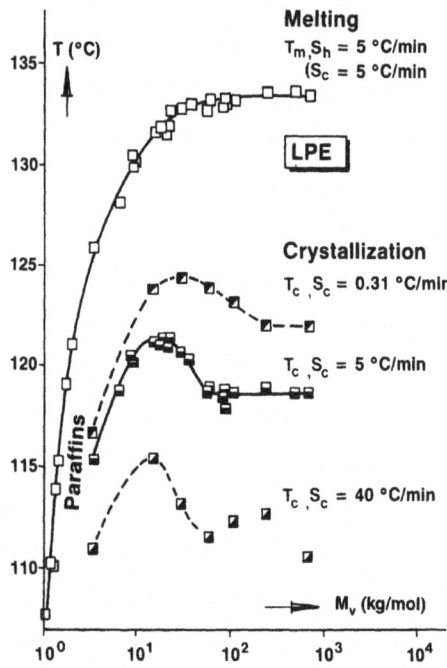

Figure 3. Crystallization and melting peak temperatures, T_c and T_m, respectively, for linear polyethylene (LPE) fractions with narrow molar mass distributions, as obtained from DSC-2 cooling and heating curves (sample masses of 0.800 ± 0.025 mg), and T_m for paraffins, as obtained from the literature, as functions of molar mass. Influence of cooling rate on T_c.

LINEAR AND BRANCHED POLYETHYLENES

Linear Polyethylene

Figure 3 shows the crystallization and melting peak temperatures, $T_c(M)$ and $T_m(M)$, respectively, for LPEs with narrow molar mass distributions as a function of M_v. Being overall quantities, these peak temperatures give only a first indication of the DSC curves for these samples. The actual curves strongly differ for the various fractions: they are narrow for low-molar-mass fractions and broad for high-molar-mass fractions.

The molar mass strongly influences T_c and T_m.[14-18] The figure shows that, at a cooling rate (S_c) of 5 °C/min, the overall crystallization rate starts to decrease at a molar mass of about 20 kg/mol and higher, resulting in a decrease in T_c. At still higher molar masses, the T_c curve levels off.

The decrease in T_c with increasing M must be due to the fact that, as the molar mass increases, it becomes increasingly difficult for the chains to be wholly or partly incorporated into lamellar crystallites at the same degree of supercooling as for the low molar mass fractions within the time frame of the measurement.[19-22] This means that with higher molar masses the degree of supercooling becomes so large that chain segments can crystallize in the same crystallite or in different crystallites, independently of one another.[17,18,23-25] All this is of course influenced by the time scale of the experiment, as can be seen from the influence of the cooling rate; see the measurements with $S_c = 0.31$ °C/min and $S_c = 40$ °C/min.

The melting peak temperatures, measured at a heating rate (S_h) of 5 °C/min after cooling at $S_c = 5$ °C min, show a different picture. The influence of molar mass is still prominent (not only for low molar masses but also for high molar masses; the difference between the experimental T_m values and the theoretical values[3,22,26,27] strongly increases with increasing M at high values of M), but a maximum of the type found in the $T_c(M)$ curves is not found.

In our opinion this illustrates an important phenomenon, viz. that during heating in the DSC at 5 °C/min after cooling at 5 °C/min reorganization/recrystallization processes occur[1,2,28] which wipe out part of the thermal history. Obviously, this casts doubt on the value of experiments in which such — very common — scanning rates are applied, since the results can no longer be interpreted in a straightforward way.

Raman longitudinal acoustic mode (LAM) measurements[29-31] provide evidence of the above-mentioned phenomenon.[28,32,33] Figure 4 is the result of calculations of the Raman length based on LAM measurements, that is, the length of an all-*trans-trans* segment of a PE chain in a lamellar crystallite (usually at an angle relative to the surface of the lamellar crystallite[34,35]), as a function of thermal history. Using the initial Raman length at nucleation, $l_R*(T_c)$, as a reference [although this $l_R*(T_c)$ is actually already larger than the real initial Raman length[36-38]] we find in practically all cases that the Raman length increases[34,39-41] during cooling as well as heating in the temperature range normally applied. The magnitude of the increase strongly depends on the temperature at which crystallization starts and on the combination of heating and cooling rate. The absolute value of l_R is of little significance here; it is the trend that is important. Clearly, considerable changes in l_R can occur due to changes in lamellar thickness.

It is also clear that these changes can have a considerable influence on the relationship between T_c and T_m, with T_c influencing the dimensions of the developing crystallites and T_m being related to the smallest dimensions of the resulting crystallites. The question remains how these effects manifest themselves in the case of different molar masses. Another important conclusion is that the results of characterizations performed at room temperature cannot always be related to T_c or T_m in a straightforward manner. This is true in particular for morphological quantities such as lamellar thickness, etc. All this implies that the usual programme of measurements at fixed — and relatively low — scanning rates is too limited to yield unambiguous relationships. One should in any case carry out measurements at various heating rates and preferably also at much higher heating rates (in the latter case heating rates should significantly exceed cooling rates!) than are customary, in order to avoid reorganization and/or recrystallization (although superheating

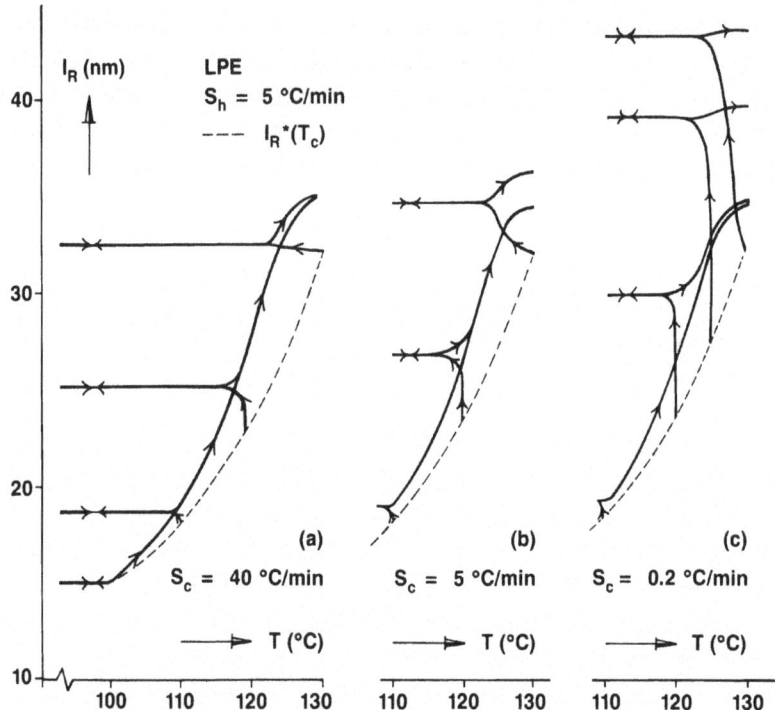

Figure 4. The Raman length ($\rightarrow$), $l_R(T)$, for LPE as a function of the Raman length at the temperature at which the second stage of crystallization starts (- - -), $l_R{}^*(T_c)$; the cooling rate [(a): $S_c = 40$ °C/min; (b): $S_c = 5$ °C/min; (c) $S_c = 0.2$ °C/min] and the subsequent heating rate ($S_h = 5$ °C/min).

should of course also be avoided). However, considering the high cooling rates applied during processing (e.g. in the film casting process), this is not feasible in a convenient way.[42-44]

Therefore, it is extremely important that morphological characterizations and DSC measurements are in any case carried out on samples with similar thermal histories; that values measured at the same temperature can be compared and that the time scales of the experiments are taken into account. The use of synchrotron sources in time-resolved X-ray measurements as a function of temperature, for example, is a major step forward.

High and Low Density Polyethylene

The $T_m(M)$ curve for LPE in Figure 3 can be used as a reference curve for polyethylenes with a low degree of short chain branching. The idea is that when a branched polymer such as HDPE is separated into fractions with narrow molar mass distributions, the influence of molar mass on the crystallization and melting behaviour of each individual HDPE fraction is the same as the influence of molar mass on the behaviour of the corresponding LPE fraction. This would mean that any differences between the $T_m(M)$ curves are ascribable to SCB.

This is confirmed by the results of measurements performed on Size Exclusion Chromatography (SEC) fractions and Direct Extraction (DE[45]) fractions of the HDPE Hizex 7000 F reported earlier[15]. Incidentally, for this polyethylene a comparison of the $T_c(M)$ curves provided the same information.

Figure 5 shows the $T_m(M)$ curves for three HDPEs.[46] The datapoints were obtained

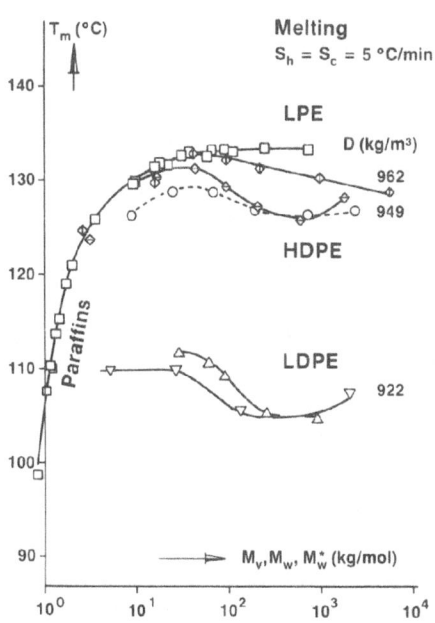

Figure 5. Melting peak temperatures for fractions ($\Phi, \bigcirc, \ominus$: "analytical" SEC) of three HDPEs and for fractions ($\triangle$: preparative SEC, sample masses of 0.800 ± 0.025 mg; $\triangledown$: "analytical" SEC) of two LDPEs from DSC-2 heating curves as functions of M_v, M_w and M_w^* for LPE, HDPE and LDPE, respectively. Densities and thermal history as indicated. LPE curve based on paraffins and a selection of LPEs.

by performing DSC measurements on fractions obtained from "analytical" SEC, used in a preparative way.[15] The densities indicated in the figure were determined at room temperature after compression moulding. The HDPE with the highest density shows higher melting peak temperatures, on average, than the HDPEs with the lowest density.

The differences in melting peak temperature distribution as a function of molar mass are important. In the case of the HDPE with the highest density, the interpretation of the data is that the SCB is incorporated into the longer chains. In the two HDPEs with the lowest density, the distributions differ essentially from one another.

Clearly, the SCB of the whole polymer is just an overall average. The same is true for the melting peak temperature of the DSC curve, although in this case the situation is more complex, because this temperature is related to the melting behaviour of a physical blend of HDPE chains, which chains obviously have demonstrably different SCBs. The density is related to the crystallization behaviour of the same blend in the same complex way.

The distribution of SCB over the chains has been found to be important for the long-term performance of polyethylene gas pipes, for the environmental stress cracking resistance of polyethylene products, etc.

The $T_c(M)$ curves for the three HDPEs (not reported here) show roughly the same behaviour, but compared with the LPE fractions the curve sections above about 10 kg/mol are partly shifted to higher temperatures: in the molar mass range between 10 and 200 kg/mol the values for the HDPE with the highest density are up to 3 degrees higher than those of LPE. Differences in nucleation density in the samples[47] might explain this to some extent; T_c is very sensitive to such differences. A remarkable finding is that here, too, such differences are completely or partly cancelled by reorganization/recrystallization during cooling and heating: the $T_m(M)$ curves for all HDPEs are below those for LPE. Therefore, any conclusions about SCB should be based on the $T_c(M)$ trend for HDPE fractions rather than on the absolute value of $T_c(M)$; better still, they should be based on the $T_m(M)$ curves.

For Hizex 7000 F, an HDPE with a density of 952 kg/m^3 after compression moulding, it has been demonstrated by DSC measurements of fractions[15] that DE gives the same separation according to M as SEC, the advantage of DE being that it yields far more material for further study. Thus, the SCB of HDPE as a function of molar mass can be determined quantitatively and relatively rapidly using DE.

The same figure shows two LDPEs which have practically the same density after compression moulding and practically the same branching content (24 and 25 CH$_3$/1000 C). This branching content includes, besides SCB, about 1 CH$_3$/1000 C due to long chain branching. To account for the influence of LCB on the molar mass distribution as determined by SEC, the molar mass values are here referred to as M* (with M* < M).

Surprisingly, $T_m(M^*)$ decreases by 6 degrees on average with increasing M*.[15,46] This cannot be due to an increase in SCB with increasing M*: SCB decreases slightly with increasing M*. $T_c(M^*)$ decreases even more strongly with increasing M*: 10 degrees on average.[15,48a] Apparently, in this case there is a qualitative similarity between the melting behaviour and the crystallization behaviour. The strong decrease in $T_c(M^*)$ and $T_m(M^*)$ with increasing M* has been observed before, and also for other polyethylenes, in roughly the same M* range.[49-56]

In the case of the LDPE fractions, $T_c(M^*)$ is highly sensitive to differences in cooling rate, whereas $T_m(M^*)$ is not. The fraction with $M_w^* = 250$ kg/mol, for example, shows an increase in T_c of 5 degrees when the cooling rate is decreased from 5 °C/min to 0.31 °C/min, while the increase in T_m upon subsequent heating at a rate of 5 °C/min is less than one degree. This is much like the situation depicted in Figure 4 for LPE for these conditions, even though SCB is likely to hinder reorganization in the case of LDPE.

Fractionation of HDPE and LDPE according to molar mass thus yields a lot of information about the molecular structure and provides us with insight into relationships between this structure and the crystallization and melting behaviour. The results also show that in a fractionation based on crystallization in solution followed by dissolving, for example Temperature Rising Elution Fractionation (TREF),[57] one should not focus on SCB only, but on the combined effects of SCB and molar mass.

This was clearly demonstrated[15,46] for the HDPE Hizex 7000 F: a fractionation according to crystallizability/dissolvability yielded some fractions with bimodal molar mass distributions, as was to be expected on the basis of the $T_m(M)$ curve. This curve showed an initial increase with M (along the LPE line) but then a strong decrease due to an increase in SCB (at 1000 kg/mol the difference with the LPE curve was more than 9 degrees; compare Figure 5). The fractions concerned include both linear short chains and long chains with SCB, which cause the bimodality in the molar mass distribution. This must of course be taken into account when studying polyethylenes that crystallize in the same range of molar masses and temperatures, for example LLDPE.

In general, especially with new polymers, a first separation according to molar mass, possibly followed by a second separation according to crystallizability/dissolvability, is to be preferred in a structure clarification study. DE followed by analytical TREF (ATREF) or preparative TREF (PTREF), by DSC or by Microcalorimetry are good combinations.[9] Only within a well-researched system might this order be reversed, for the purpose of comparing samples within such a system.[57-63] One should be particularly careful when using ATREF alone, as is often done in practice. This may yield too little information to justify the time-consuming procedure; the use of quantitative DSC on pure polymer or microcalorimetry on polymer plus solvent[9] might very well have produced the same information.

HOMOGENEOUS AND HETEROGENEOUS ETHYLENE COPOLYMERS

Homogeneous and Heterogeneous Ethylene-1-Octene Copolymers

We shall now discuss polyethylenes with such a large number of comonomer units incorporated into the chains that the resulting SCB makes crystallization considerably more difficult than in the case of LPE and HDPE. One of the effects is a substantial decrease in overall density; the ethylene copolymers to be discussed have densities between, roughly, 920 and 855 kg/m³, see Figure 1.

Figure 6 shows two ethylene-1-octene copolymers[64] that currently represent two extremes as regards molecular structure. They both have about the same octene content, X_8, but they differ very much in crystallization and melting behaviour. The EO copolymer crystallizes below about 100 °C and has a long crystallization range, which extends into the glass transition region around -60 °C. Its melting range is correspondingly wide. The VLDPE has an even wider crystallization range, which starts at a higher point, viz. around

120 °C. With this sample, melting starts in the glass transition region at about -60 °C and continues all the way up to the endpoint of about 130 °C. The DSC curves of the EO copolymer are single-peaked, whereas those of the VLDPE are double-peaked, see the classification in Figure 1 (the peaks referred to here are the crystallization peak at about 70 °C for the EO copolymer and the crystallization peaks at about 70 and 109 °C for the VLDPE; the melting peak temperatures are 76 °C and 91 and 121/125 °C, respectively). Another remarkable observation is that the areas under the curves differ widely. The EO copolymer clearly has the lowest crystallinity at room temperature, and therefore the lowest density: another way of saying this is that in this sample comonomer incorporation has very effectively reduced the density. The material properties are of course also different.

The broad crystallization and melting ranges are caused by the distribution of crystallizable ethylene sequences in the chains (ESLD), which is related to the way in which the comonomer units were added to the chain. If the ethylene sequences have lengths of less than a few hundred, a difference in ethylene sequence length (ESL) will result in a difference in crystallization and melting temperature, analogous to the influence of molar mass on T_c and T_m for M values less than about 10 kg/mol, see Figure 3. In absolute terms, the effects of ESL on T_c and T_m will differ from the effects of M, because

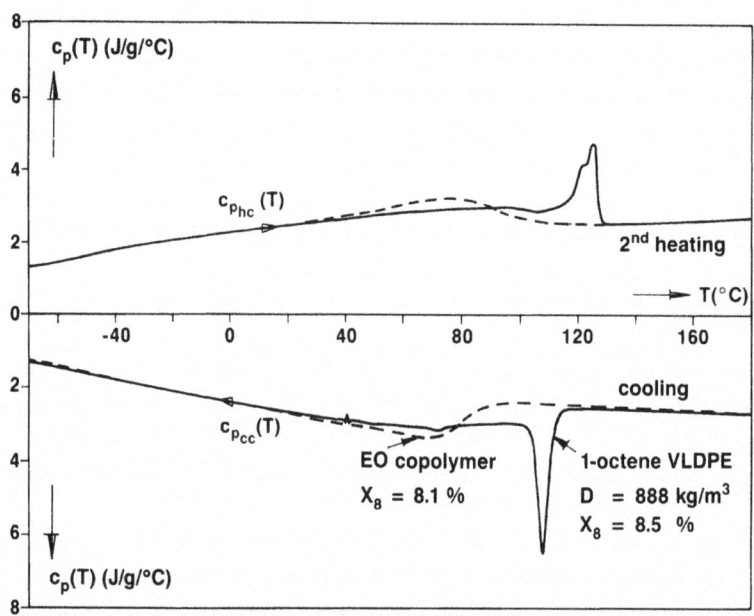

Figure 6. DSC-2 continuous specific heat capacity curves for cooling, $c_{p_{cc}}(T)$, at a rate of 10 °C/min, from 180 °C to -70 °C and subsequent heating, $c_{p_{hc}}(T)$, at a rate of 10 °C/min, for an EO copolymer and a VLDPE. Isothermal stays of 5 min, sample masses of 9.259 mg (EO) and 10.005 mg (VLDPE).

short chains obviously crystallize and melt in a way that differs from ethylene sequences of the same length incorporated into longer chains.

In addition, the two copolymers apparently differ strongly in ESLD: the differences in crystallization and melting range are too big to be explained by differences in molar mass distribution.

A useful classification of copolymers,[9,65] which dates a long way back,[66] is the division into homogeneous and heterogeneous types (see Figure 1). For copolymers the terms "homogeneous" and "heterogeneous" will be defined as follows with respect to comonomer incorporation:

- Homogeneous copolymer: we speak of a homogeneous copolymer when the way the comonomer is incorporated during polymerization can be described by one single set of chain propagation probabilities of (co)monomer addition to the chain (P set) or, alternatively, by the combination of a single set of reactivity ratios (r set) and a single monomer/comonomer ratio. Statistically there are no differences within and between the chains.
- Heterogeneous copolymer: all other copolymers are heterogeneous. Two special cases are distinguished, viz. "intramolecular heterogeneity" of comonomer incorporation when the heterogeneity is manifested within the chains, and "intermolecular heterogeneity" when the heterogeneity is between chains.

The "homogeneous"/"heterogeneous" dichotomy need not coincide with "single"-peaked/"multiple"-peaked DSC curves: according to the definition just given, HDPE is heterogeneous (see Figure 5!), although its DSC curve is single-peaked. LDPE is heterogeneous, because due to the branching mechanism numerous different short and long side branches can occur with very complex distributions among the chains, although here, too, the DSC curve is single-peaked.

The catalyst used in the preparation of the EO copolymer in Figure 6, the polymerization method and analytical data on the sample all make this copolymer homogeneous according to the definition given.[67] The VLDPE in the same figure is heterogeneous.[64] In the rest of this chapter we shall elaborate on the homogeneous/heterogeneous classification with reference to examples.

Homogeneous Ethylene-Propylene Copolymers with Inversion

Earlier we indicated that it is difficult to determine the ESLD for a copolymer and that, moreover, no quantitative models are available to translate an ESLD at a chosen thermal history to, for example, a DSC curve.

For a range of homogeneous ethylene-propylene copolymers it was possible to determine the ESLDs in detail.[68-72] For this we had to develop a terpolymer model, on account of the fact that with the — promoted — catalyst system used (consisting of an aluminium alkyl combined with a vanadium component) the propylene unit adds to the chain in a "normal" or "inverted" position. The model is capable of explicitly determining the degree of inversion and thus the reactivities associated with inversion, alongside the inversion-independent reactivities.[69,72] The experimental basis of the model is formed by the relative amounts by mass of methylene in sequences with lengths of 1 through 5 separately, the pooled sequences of length 6 and longer and the mass fraction of propylene, as determined by ^{13}C-NMR, plus the ethylene-propylene molar ratio in the reactor.

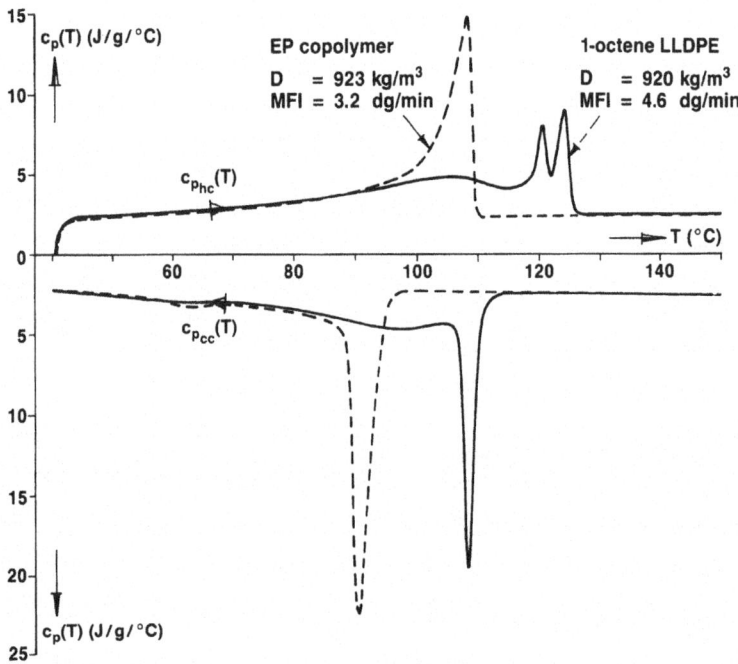

Figure 7. DSC-2 continuous specific heat capacity curves for cooling, $c_{p_{cc}}(T)$, at a rate of 5 °C/min, from 200 °C to 40 °C and subsequent heating, $c_{p_{hc}}(T)$, at a rate of 5 °C/min. Isothermal stays of 5 min, sample masses of 4.870 mg (EP) and 4.976 mg (LLDPE).

    The model was successfully used to analyse (homogeneous) EP(DM)s, but it can be applied more generally to copolymers in which one of the units can be added to the chain in an inverted position and for which NMR can provide the required data. Of course, a copolymer in which inversion does not occur is a special case in the model.

    In Figure 7 a homogeneous EP copolymer from the series investigated is compared with a heterogeneous LLDPE.[73] Although the densities are the same, and therefore the areas under the curves and the crystallinities at room temperature are also the same, the difference in the distribution of crystallization and melting temperatures is unmistakable.

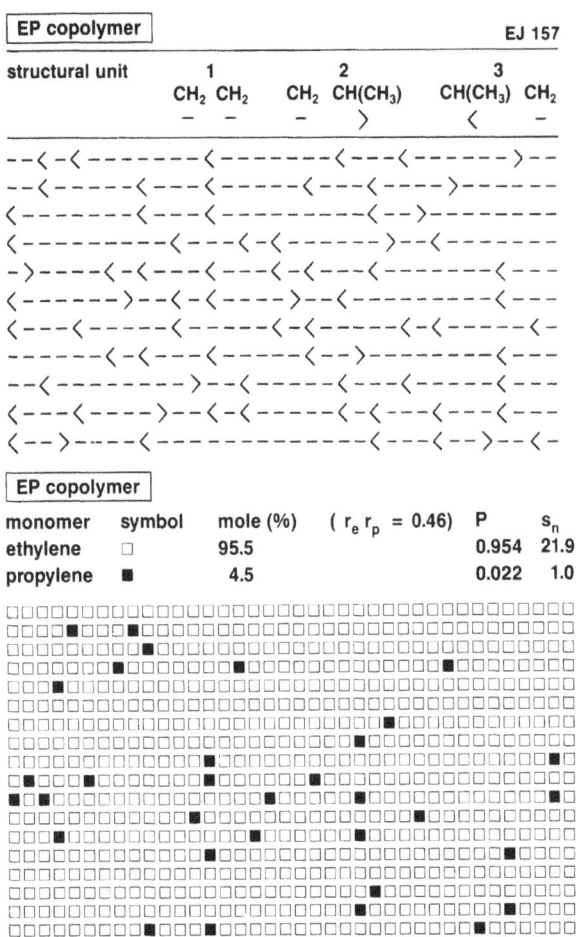

Figure 8. (upper part) Simulation of the chain structure of a homogeneous EP copolymer with $X_c = 62\%$ and a percentage of inversion of 19%. Terpolymer (- - ethylene; $<$ - normal propylene; - $>$ inverted propylene) presentation. The chain is constructed by linking the end of each line to the beginning of the next line.   (lower part) Simulation of the homogeneous EP copolymer, $X_c = 95.5\%$, shown in Figure 7. Copolymer presentation (ethylene and propylene). $P_{ee} = 0.954$ and $P_{pp} = 0.022$ are chain propagation probabilities; s_n is the number-average sequence length. Chain construction as in the upper part.

These differences in distribution are related to differences in crystallite dimensions and, ultimately, differences in ESLD.

Figure 8 (lower part) gives an impression[9] of the chain structure (the lines should be read from left to right as in a book) of the EP copolymer from Figure 7, with an ethylene mole percentage, X_e, of 95.5%, consisting of ethylene units and propylene units. In this representation no distinction is made between "normal" and "inverted" propylene. This picture is part of the result of a complete Monte Carlo simulation of the chain structure; the upper part of Figure 8 shows part of a similar simulation[69] for a different homogeneous (crystallizable) EP copolymer. Here, the difference between normal (indicated by the number 3) and inverted (indicated by the number 2) propylene is indicated. Obviously, the propylene units will hinder the crystallization process greatly, and the ethylene sequences with their length and specific location in the chains will be the determining structural entities during crystallization. The resulting distributions, such as methylene, ethylene and propylene distributions, can be calculated directly from the reactivity ratios and the ethylene/propylene ratio in the reactor. The following r set[69] applies to the series of EP copolymers discussed here: $r_{11} = 1$; $r_{12} = 97$; $r_{13} = 19.7$; $r_{21} = 0.0103$; $r_{22} = 1$; $r_{23} = \infty$; $r_{31} = 0.0278$; $r_{32} = 2.7$; $r_{33} = 1$. For the copolymers discussed here, homogeneity means that in each chain, whatever its length, the same statistics apply with regard to ethylene and propylene additions, viz. the statistics defined by the r set.

The chain structures of these EP copolymers are thus well known, but it is not easy to relate these structures to the crystallization and melting behaviour of the polymers. This is one of the subjects into which research is currently being carried out.[48b] For research in this field, information of the same detailed nature as given above for these EP copolymers needs to become available for other homogeneous ethylene copolymers as well (such as the EO copolymer depicted in Figure 6).[67]

Heterogeneous Ethylene-1-Octene Copolymers: Linear Low Density and Very Low Density Polyethylenes

Copolymers such as the VLDPE in Figure 6 and the LLDPE in Figure 7 can be heterogeneous for a variety of reasons, for example because the catalyst used has more than one active site,[66,74-78] the ethylene/comonomer ratio is not constant during polymerization, or several reactors are used. If, in extreme cases, the resulting ESLD is multi-peaked, as is to be expected with heterogeneous LLDPE and VLDPE, the crystallization process will take place in different temperature ranges, resulting in crystallites of different dimensions and a corresponding, possibly multi-peaked, size distribution. This size distribution, in its turn, will give rise to a melting point distribution, which may be multi-peaked. For some catalyst systems this chain of causes and effects is fairly straightforward, which means that in such cases the crystallization and melting behaviour contains clues to the molecular structure.

Figure 9 illustrates the multi-peak character of the excess specific heat capacity for a VLDPE with a density after compression moulding of 899 kg/m^3 and for which X_8 is 6.6%.[79] The excess c_p is the difference between the c_p and the contribution from the base-line c_p. To obtain this excess c_p, the experimental $c_p(T)$ was compared with the $c_{pa}(T)$ for completely amorphous PE and the $c_{pc}(T)$ for completely crystalline PE[80,81] (using the two-phase model, and assuming octene to be excluded from the crystals). The two-phase model was used because a comparison of c_p measurements and density measurements had shown there was no reason to evaluate the results using a three-phase model.[8,64] In the melt, the excess c_p should be zero, but this is not quite true in Figure 9 because this particular

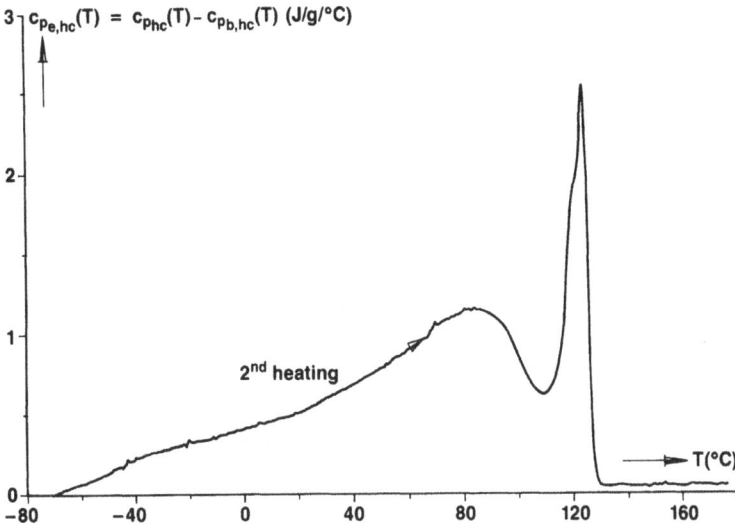

Figure 9. The excess-c_p for the heating curve, $c_{pe,hc}(T)$, as a result of subtraction of the base-line c_p for the heating curve, $c_{pb,hc}(T)$, from the specific heat capacity curve for heating, $c_{phc}(T)$. $c_{phc}(T)$ is based on DSC-2 continuous heating curves at 10 °C/min after cooling at 10 °C/min from 177 °C to -73 °C for an empty pan plus lid and exactly the same procedure for the same pan and lid with the VLDPE sample (10.768 mg) in it.

measurement in this particular range was not entirely quantitative. It is the excess function that is of interest for correlations with the molecular structure and which forms the basis of a quantitative comparison with the lamellar thickness distribution and, ultimately, with the ESLD. Incidentally, this excess function can also be obtained in a different way, that is, other than via a c_p measurement.[8,79]

It is impossible to gain a true understanding of the molecular structure of this type of heterogeneous copolymers without using fractionation techniques. As mentioned before, a first fractionation according to molar mass is to be preferred.

Figure 10 shows the result of direct extractions of two 1-octene LLDPEs with the same density at room temperature after compression moulding and roughly the same DSC curves; the DSC curve of the LLDPE with MFI = 4.6 dg/min is given in Figure 7. The MFIs are widely different due to the widely different molar mass distributions. The average molar masses (as determined with the aid of SEC) for the LLDPE with MFI = 4.6 dg/min are $M_n^* = 20$, $M_w^* = 80$ and $M_z^* = 300$ kg/mol, and those for the LLDPE with MFI = 0.1 dg/min are $M_n^* = 21$, $M_w^* = 280$ and $M_z^* = 1200$ kg/mol. For the latter polymer the distribution is bimodal. Clearly, the distribution of 1-octene comonomer units over the chains can differ greatly, as was observed earlier with some 1-butene LLDPEs.[65]

It is also clear that with this type of copolymer, whose molecular structure can be very complex, various analytical techniques have to be used: in the example given here,

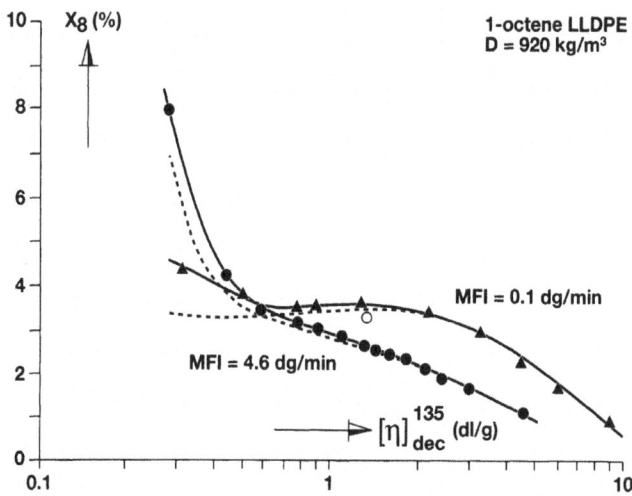

Figure 10. Mole percentages of 1-octene comonomer, X_8 (%), based on total $CH_3/1000$ C, and intrinsic viscosities, $[\eta]$ (dl/g), for 2 LLDPEs: whole polymer ($\bigcirc$) and direct extraction fractions ($\bullet$, $\blacktriangle$). - - - after first-order correction of X_8 for endgroups.

the only information provided by DSC is that the copolymers are heterogeneous. Incidentally, in the case of heterogeneous LLDPEs a fractionation according to molar mass, as was done here with the aid of DE, would provide only half of the picture. The DSC curves, see Figure 9 in ref. 65, of the LLDPE with MFI = 4.6 dg/min are still broad, with melting regions going roughly from room temperature to 130 °C: the fractions are still heterogeneous.

Although SCB quite often increases on average with decreasing $[\eta]$ (or M),[65] see Figure 10 for the LLDPE with MFI = 4.6 dg/min, it is apparently not true that only the short chains have a high degree of SCB and the long chains have no SCB. An extremely important conclusion from the fact that the DSC curves for the DE fractions are still broad is that all fractions — each of which has a narrow molar mass distribution — still consist of (equally long) chains ranging from unbranched to heavily branched (intermolecular heterogeneity) or, alternatively, consist of equally long chains with segments ranging from unbranched to heavily branched (intramolecular heterogeneity). The difference between the fractions is thus mainly a difference in average amount of branching. The exact relationship found between X_8 and $[\eta]$ (or M) depends on the fractionation method used, because every fractionation method yields a different cross-section of the three-dimensional distribution W - X_8 - $[\eta]$, where W is the relative frequency of occurrence of a chain with a specific X_8 - $[\eta]$ combination. In the case of preparative SEC (PSEC), for example, the line through the averages in the X_8 - $[\eta]$ plane can be more parallel with the $[\eta]$ axis than in the case of DE, because with DE the separation according to molecular dimension slightly deteriorates as the (short) side chains become longer.[82]

Let us now look at a different cross-section of the three-dimensional distribution, viz. a separation based on crystallizability/dissolvability.[83] The purpose of such a fractionation is to determine whether a split can be made between chains with a lot of SCB and equally long chains with little SCB. If the split is possible, there is intermolecular heterogeneity. If it is not, intramolecular heterogeneity may be the cause. A simple way of establishing intermolecular heterogeneity is a DSC measurement of the fractions.

Figure 11 shows the result of such a fractionation[64] of the VLDPE from Figure 6.

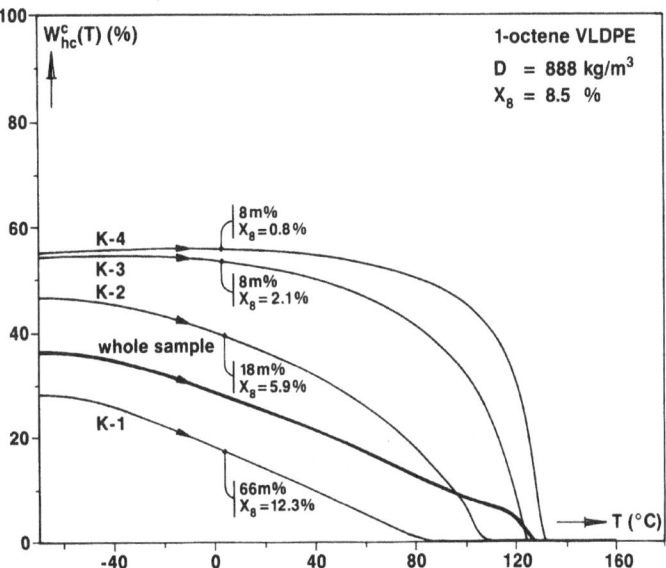

Figure 11. Enthalpy-based mass crystallinity heating curves, $W^c_{hc}(T)$, for the VLDPE shown in Figure 6, based on heat capacity curves for fractions obtained via a crystallization/dissolution method.

The material was dissolved in xylene, then crystallized in a stirred solution at a rate of less than 1 °C/min and subsequently heated. The fractions were drained off, first at room temperature (K-1) and then, after heating with stirring, at 68 °C (K-2), 91 °C (K-3), and finally at 110 °C (K-4). The crystallinity curves as a function of temperature, calculated from the c_p curves of the fractions using the two-phase model,[8,80,81] show a large difference in melting behaviour. Fraction K-1 shows no melting at the highest temperatures, which means it contains little or no unbranched material. Fraction K-4, on the other hand, shows no melting at low temperatures, which means it contains no strongly branched material.

The fact that a physical split into such fractions is possible indicates intermolecular heterogeneity. If we combine this with DE results,[64] it is clear that the VLDPE in question is a physical blend of chains, with chains of the same length showing large differences in SCB, and the short chains being branched more strongly, on average, than the longer ones. The same conclusion was drawn earlier for heterogeneous LLDPEs.[65] The situation can of course differ from one catalyst system to another and from one polymerization process to another, and therefore the conclusion regarding the intermolecular character of the heterogeneity cannot be generalized.

Incidentally, at room temperature 66% (m/m) of the VLDPE is in solution, see fraction K-1 in Figure 11, and this shows that with this type of copolymers it is technically difficult to properly fractionate according to crystallizability/dissolvability. A DE fractionation of this polymer presented no problems.

Because of the large differences in comonomer content between the various chains of the VLDPE one might expect the morphology, too, to reveal a blend character, for the differences in crystallization temperature are so big that, besides some co-crystallization, there must be differences in crystallite dimensions.

Figure 12 shows a transmission electron micrograph of a clustering of lamellar crystallites into nodular entities, which we have termed compact semi-crystalline domains, CSDs, in a seemingly structureless matrix.[84]   In the CSDs long and short lamellae are

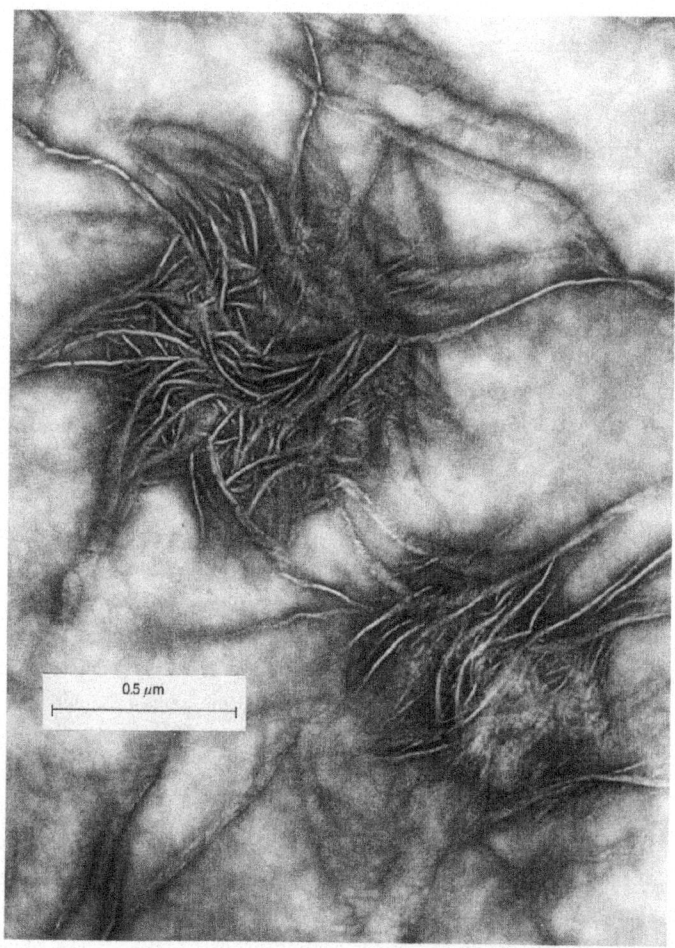

Figure 12. Morphology as revealed via TEM of the VLDPE of Figs. 6 and 11, after staining/fixation via intensive chlorosulphonation. Compact semi-crystalline domains (CSDs) with lamellae inside and between them.

visible which differ in thickness. Higher magnifications reveal the presence of very small lamellar crystallites that have crystallized on to the larger ones. The internal morphology of a CSD resembles that of LLDPE. The CSDs are interconnected by lamellae, resulting in a network in which the CSDs are nodes. Such networks have also been found to exist in heterogeneous VLDPEs with higher densities. The long lamellae no doubt consist of chains giving rise to the highest crystallization and melting peak in the DSC curve: HDPE-like chains present in the blend (which differ in length). The intermediate and shorter lamellae probably correspond to more heavily branched chains which give rise to the lowest crystallization and melting peak in the DSC curve, see the discussion at Figure 6.

There are three reasons why the matrix looks structureless. First of all, staining through chlorosulphonation[85] in the vapour phase[86] to obtain the required contrast in the electron microscope took place at 45 °C, so we know that at that temperature about 80% (m/m) of the material (the matrix) was in the molten state, see the enthalpy-based mass crystallinity curve for the VLDPE in Figure 11. Secondly, a more severe fixation/staining treatment was applied to reveal the internal morphology of the CSDs, and this treatment can affect the dimensions of morphological features, such as lamellar thickness. Thirdly, crystals smaller than 3 to 4 nm cannot be made visible by the fixation/staining technique used. Thus, at room temperature the matrix will be a phase of very low crystallinity, composed of heavily branched chains which crystallize with difficulty, if at all, and which are therefore situated in the amorphous regions or in crystallites with small dimensions. Note that here, too, technical problems arise, namely when one attempts to perform staining/fixation (to study the morphology) at lower temperatures.

Thus we find for this VLDPE a co-continuous structure, while it seems that the network structure consisting of interconnected CSDs occurs more generally in VLDPEs. The disperse phase, such as the CSDs (but also the disperse low-crystalline phase observed in VLDPEs with higher densities) occurs in smooth, round domains, which points to partial thermodynamic demixing in the melt of species which differ strongly in branching content.[87-90] The general conclusion is that in this type of copolymers there is a close relationship between morphology and molecular structure.

For a complete picture of the molecular structure a combination of a separation according to chain length (for example via DE or SEC) and a separation according to ESL (for example via TREF) is necessary.[10,61,65,91-94]

Figure 13 shows the results of a cross-fractionation of a 1-butene LLDPE.[46] The first fractionation, via PSEC, was according to chain length. The five fractions thus obtained were fractionated via ATREF. A fractionation via SEC is actually a fractionation according to molecular dimension, but this can be unambiguously translated to molar mass and chain length if the type and amount of comonomer are known.[95]

Five grams of material were used for the fractionation via PSEC [using in-house built equipment, with 1,2,4,-trichlorobenzene (TCB) as solvent, at 140 °C]. The distribution along the log M axis in Figure 13 is discontinuous because the sample was fractionated preparatively. The first and last values of log M at each "M-slice" correspond to the beginning and end of the period in which the fraction concerned was taken. The volume of each slice was normalized on the basis of the relative mass of each PSEC

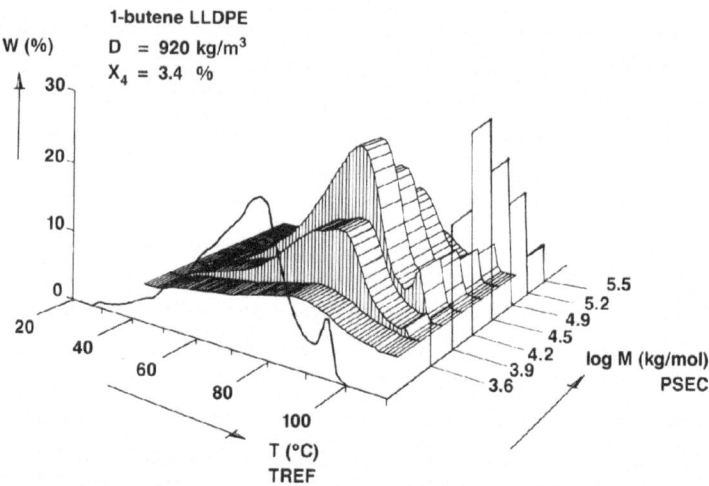

Figure 13. Cross-fractionation of a 1-butene LLDPE via analytical temperature rising elution fractionation after preparative size exclusion chromatography. In addition, the mathematical sums of the separate values for the fractions are indicated in two directions (arbitrary scales).

fraction. The final relative mass, W, is indicated on the z axis.

For cross-fractionation via ATREF 150-mg fractions of worked-up polymer, or fractions containing as much material as possible, were dissolved in 10 ml of TCB at 140 °C and cooled to 20 °C in 80 hours. After storing in a refrigerator at 0 °C the polymer was eluted using 0.5 ml of TCB/min starting at 20 °C, while the temperature was increased by 20 °C/hour.

The HDPE-like material ("high"-temperature TREF peak) is relatively prominent at higher molar masses, while its dissolution temperature is practically molar-mass-independent. The "low"-temperature TREF peak shifts to lower dissolution temperatures with decreasing molar mass. Moreover, the TREF curves appear to become slightly narrower with increasing molar mass. Another striking fact is that the "low"-temperature peak obtained in ATREF (which corresponds to the "low"-temperature peak of the DSC curve) is found to represent much more material than the "high"-temperature peak obtained in ATREF (which corresponds to the "high"-temperature peak of the DSC curve), whereas from the dq/dTs of the DSC curves of the pure polymer the amounts seem to be more balanced at first sight.

The order of fractionation applied here, first PSEC (or, as an alternative, DE) and then ATREF (or PTREF), is a deliberate choice: in view of the influence of both chain length and SCB on crystallization — see the remarks made in the sections on LPE, HDPE and LDPE — the preferred order is a fractionation according to chain length/molecular dimension and then according to SCB/ESL via crystallizability/dissolvability fractionation.

Finally, we have a comment to make about intramolecularly heterogeneous copolymers. When such a copolymer is analysed with TREF, the resulting curve will be relatively narrow. After all, the longer crystallized sequences in a chain will have a dominant influence on the dissolution of the chain and hence on the detection of this process. On the other hand, a microcalorimeter, operating under the same conditions, can record the heat effect of dissolution of all crystallized sequences — including the shorter ones — and will therefore give a relatively wide curve. Microcalorimetry is thus expected to prove very useful for specialists who want to obtain a better understanding of what is happening under specific fractionating conditions: with the aid of a microcalorimeter they can perform crystallization and dissolution measurements under these conditions.[9]

ACKNOWLEDGEMENT

The author wishes to express his appreciation to L. Coosemans for the polymerizations; to M. Pijpers for most of the DSC measurements and the crystallization/dissolution fractionation; to N. Meijerink and coworkers for the analytical SEC fractionations and SEC measurements; to A. Brands, W. Bunge and H. Schoffeleers for direct extraction, preparative SEC fractionation, TREF and viscosimetry; to A. Pijpers for the transmission electron micrograph; to A. Veermans for IR; and to H. Rhebergen, who provided the English translation of the original Dutch text.

REFERENCES

1. Wunderlich, B.: "Macromolecular Physics, Vol. 1: Crystal Structure, Morphology, Defects", Academic Press, New York (1973).
2. Wunderlich, B.: "Macromolecular Physics, Vol. 2: Crystal Nucleation, Growth, Annealing", Academic Press, New York (1976).
3. Wunderlich, B.: "Macromolecular Physics, Vol. 3: Crystal Melting", Academic Press, New York (1980).
4. Dosière, M., in: "Handbook of Polymer Science and Technology", Vol. 2, Cheremisinoff, N.P. (Ed.), Marcel Dekker, New York, 367 (1989).
5. Mandelkern, L. in: "Comprehensive Polymer Science", Vol. 2: "Polymer Properties", Booth, C., Price, C. (Eds.), Pergamon Press, Oxford, 363 (1989).
6. Phillips, P.J., Rep. Prog. Phys. 53, 549 (1990).
7. Armitstead, K., Goldbeck-Wood, G., Adv. Polym. Sci. 100, 219 (1992).
8. Mathot, V.B.F., "Thermal Characterization of States of Matter" in: "Calorimetry and Thermal Analysis on Polymers", Mathot, V.B.F. (Ed.), Hanser Publishers, 1993.
9. Mathot, V.B.F., "The Crystallization and Melting Region" in: "Calorimetry and Thermal Analysis on Polymers", Mathot, V.B.F. (Ed.), Hanser Publishers, 1993.
10. Alamo, R., Domszy, R., Mandelkern, L., J. Phys. Chem. 88, 6587 (1984).
11. Alamo, R.G., Mandelkern, L., Macromolecules 24, 6480 (1991).
12. Ergoz, E., Fatou, J.G., Mandelkern, L., Macromolecules 5, 147 (1972).
13. Mandelkern, L., Glotin, M., Benson, R.A., Macromolecules 14, 22 (1981).
14. Barrales-Rienda, J.M., Fatou, J.M.G., Polymer 13, 407 (1972).
15. Mathot, V.B.F., Pijpers, M.F.J., Polymer Bulletin 11, 297 (1984).
16. Magill, J.H., in: "Polymer Handbook" 3rd ed., Brandrup, J., Immergut, E.H. (Eds.), Wiley, VI/279 (1989).

17. Fatou, J.G., Marco, C., Mandelkern, L., Polymer, 31, 890 (1990).

18. Fatou, J.G., Marco, C., Mandelkern, L., Polymer, 31, 1685 (1990).

19. de Gennes, P.G., J. Chem. Phys. 55, 572 (1971).

20. Disc. Faraday Soc. 68: "Organization of Macromolecules in the Condensed Phase" (1979).

21. Hoffman, J.D., Polymer 23, 656 (1982).

22. Hoffman, J.D., Miller, R.L., Macromolecules 21, 3038 (1988).

23. Phillips, P.J., Polym. Prepr. (Am. Chem. Soc., Div. Polym. Chem.) 20(2), 483 (1979).

24. Hoffman, J.D., Polymer, 24, 3 (1983).

25. Hoffman, J.D., Miller, R.L., Macromolecules 22, 3038 (1989).

26. Mandelkern, L., Stack, G.M., Macromolecules 17(4), 871 (1984).

27. Mandelkern, L., Prasad, A., Alamo, R.G., Stack, G.M., Macromolecules 23, 3696 (1990).

28. Chivers, R.A., Barham, P.J., Martinez-Salazar, J., Keller, A., J. Polym. Sci.: Polym. Phys. Ed. 20, 1717 (1982).

29. Snyder, R.G., Krause, S.J., Scherer, J.R., J. Polym. Sci.: Polym. Phys. Ed. 16, 1593 (1978).

30. Snyder, R.G., Scherer, J.R., J. Polym. Sci.: Polym. Phys. Ed. 18, 1421 (1980).

31. Glotin, M, Mandelkern, L., J. Polym. Sci.: Polym. Phys. Ed. 21, 29 (1983).

32. Barham, P.J., Chivers, R.A., Jarvis, D.A., Martinez-Salazar, J., Keller, A., J. Polym. Sci.: Polym. Lett. Ed. 19(11), 539 (1981).

33. Barham, P.J., Jarvis, D.A., Keller, A., J. Polym. Sci.: Polym. Phys. Ed. 20, 1733 (1982).

34. Voigt-Martin, I.G., Adv. Polym. Sci. 67, 194 (1985).

35. Voigt-Martin, I.G., Mandelkern, L., J. Polym. Sci.: Part B: Polym. Phys. 27, 967 (1989).

36. Martinez-Salazar, J., Barham, P.J., Keller, A., J. Mater. Sci. 20, 1616 (1985).

37. Barham, P.J., Chivers, R.A., Keller, A., Martinez-Salazar, J., Organ, S.J., J. Mater. Sci. 20, 1625 (1985).

38. Barham, P.J., Keller, A., J. Polym. Sci.: Part B: Polym. Phys. 27, 1029 (1989).

39. Ungar, G., Keller, A., Polymer 27, 1835 (1986).

40. Noid, D.W., Sumpter, B.G., Wunderlich, B., Polym. Commun. 31, 304 (1990).

41. Hikosaka, M., Polymer 31, 458 (1990).

42. Hager, Jr., N.E., Rev. Sci. Instrum. 35(5), 618 (1964).

43. Hager, Jr., N.E., Rev. Sci. Instrum. 43(8), 1116 (1972).

44. Wu, Z.Q., Dann, V.L., Cheng, S.Z,D., Wunderlich, B., J. Thermal Anal. 34, 105 (1988).

45. Holtrup, W., Makromol. Chem. 178, 2335 (1977).

46. Mathot, V., Pijpers, T., Bunge, W., in: Polym. Prepr. (Am. Chem. Soc., Div. Polym. Chem.), ACS Meeting "Recent Advances in Polyolefin Polymers", Washington, 143 (1992).

47. Narh, K.A., Odell, J.A., Keller, A., Fraser, G.V., J. Mater. Sci. 15(8), 2001 (1980).

48. (a) Mathot, V.B.F. in: "Crystallization of Polymers", Dosière, M. (Ed.), NATO ASI-C Series "Mathematical and Physical Sciences", (1993).
    (b) Van Ruiten, J., Van Dieren, F., Mathot, V.B.F., in: "Crystallization of Polymers", Dosière, M. (Ed.), NATO ASI-C Series "Mathematical and Physical Sciences", (1993).

49. Murata, K., Kobayashi, S., Kobunshi Kagaku 26, 536 (1969)

50. Shirayama, K., Kita, S., Kobunshi Kagaku 28, 321 (1971).

51. Hosoi, M., Naoi, T., Kawai, T., Kuriyama, I., Kobunshi Kagaku 29, 557 (1972). Eng. Ed. 1, 848 (1972).

52. Otocka, E.P., Roe, R.J., Bair, H.E., J. Polym. Sci., Polym. Phys. Ed. 12, 1245 (1974).

53. Hser, J.-H., Carr, S.H., Polym. Eng. Sci. 19, 436 (1979).

54. Gianotti, G., Cicutta, A., Romanini, D., Polymer 21, 1087 (1980).

55. Alamo, R.G., Mandelkern, L., Macromolecules 22, 1273 (1989).

56. Alamo, R.G., Chan, E.K.M., Mandelkern, L., Voigt-Martin, I.G., Macromolecules 25(24), 6381 (1992).

57. Wild, L., Adv. Polym. Sci. 98, 1 (1990). Wild, L., Blatz, C., in: Polym. Prepr. (Am. Chem.

Soc., Div. Polym. Chem.), ACS Meeting "Recent Advances in Polyolefin Polymers", Washington, 153 (1992).

58. Nakano, S., Gotoh, Y., J. Appl. Polym. Sci. 26, 4217 (1981).

59. Brauer, E., Gebauer, E., Wiegleb, H., Fuerling, W., 31st IUPAC Macromol. Symp. (MACRO'87), Merseburg, V/Po/28 (1987).

60. Mirabella, F.M., Jr., Ford, E.A., J. Polym. Sci.: Part B: Polym. Phys. 25, 777 (1987).

61. Bodor, G., Dalcomo, H.J., Schröter, O., Colloid & Polymer Sci. 267, 480 (1989).

62. Hazlitt, L.A., Moldovan, D.G., US Pat. No. 4,798,081 (1989).

63. Karbashewski, E., Kale, L., Rudin, A., Tchir, W.J., Cook, D.G., Pronovost, J.O., J. Appl. Polym. Sci. 44, 425 (1992).

64. Mathot, V.B.F., Pijpers, M.F.J., J. Appl. Polym. Sci. 39(4), 979 (1990).

65. Mathot, V., in "Polycon '84 LLDPE", The Plastics and Rubber Institute, London, 1 (1984).

66. Elston, C.T., Can. Pat. No. 984,213 (1967).

67. Hunter, B.K., Russell, K.E., Scammell, M.V., Thompson, S.L., J. Polym. Sci.: Polym. Chem. Ed. 22, 1383 (1984). Clas, S.-D., McFaddin, K.E., Russell, K.E., Scammel-Bullock, M.V., Peat, I.R., J. Polym. Sci.: Part A: Polym. Chem. 25, 3105 (1987).

68. Mathot, V., Pijpers, M., Beulen, J., Graff, R., van der Velden, G., in: "Proceedings of the Second European Symposium on Thermal Analysis 1981 (ESTA-2)", Dollimore, D. (Ed.), Heyden, London, 264 (1981).

69. Mathot, V.B.F., Fabrie, Ch.C.M., Tiemersma-Thoone, G.P.J.M., van der Velden, G.P.M., in: "Proceedings Int. Rubber Conf. (IRC)", Kyoto, October 15-18, 1985, 334 (1985).

70. Mathot, V.B.F., Fabrie, Ch.C.M., J. Polym. Sci.: Part B: Polym. Phys. 28, 2487 (1990).

71. Mathot, V.B.F., Fabrie, Ch.C.M., Tiemersma-Thoone, G.P.J.M., van der Velden, G.P.M., J. Polym. Sci.: Part B: Polym. Phys. 28, 2509 (1990).

72. Mathot, V.B.F., Fabrie, Ch.C.M., Tiemersma-Thoone, G.P.J.M., van der Velden, G.P.M., to be published.

73. Mathot, V.B.F., Pijpers, M.F.J., Thermochim. Acta 93, 3 (1985).

74. Cooper, W., in: "Comprehensive Chemical Kinetics 15, Non-Radical Polymerization", Bamford, C.H., Tipper, C.F.H. (Eds.), Elsevier, Amsterdam, Ch. 3, 133 (1976).

75. Kissin, Y.V.: "Isospecific Polymerization of Olefins with Heterogeneous Ziegler-Natta Catalysts", Springer Verlag, New York (1985).

76. Cozewith, C., Macromolecules 20, 1237 (1987).

77. McAuley, K.B., MacGregor, J.F., Hamielec, A.E., AIChE J. 36(6), 837 (1990).

78. Cheng, H.N., Macromolecules 25, 2351 (1992).

79. Mathot, V.B.F., Pijpers, M.F.J., Thermochim. Acta 151, 241 (1989).

80. Mathot, V.B.F., Pijpers, M.F.J., J. Therm. Anal. 28, 349 (1983).

81. Mathot, V.B.F., Polymer 25, 579 (1984). Errata: Polymer 27, 969 (1986).

82. Bunge, W., Tacx, J., unpublished results.

83. Koningsveld, R., Kleintjens, L.A., Geerissen, H., Schützzeichel, P., Wolf, B.A. in: "Comprehensive Polymer Science", Vol 1: "Polymer Characterization", Booth, C., Price, C. (Eds.), Pergamon Press, Oxford, 293 (1989).

84. Deblieck, R.A.C., Mathot, V.B.F., J. Mater. Sci. Lett. 7, 1276 (1988).

85. Kanig, G., Colloid & Polymer Sci. 260(4), 356 (1982).

86. von Bassewitz, K., zur Nedden, K., Kautschuk Gummi Kunsstoffe 38, 42 (1985).

87. Hill, M.J., Barham, P.J., Keller, A., Rosney, C.C.A., Polymer 32(8), 1384 (1991).

88. van Ruiten, J., Boode, J.W., Polymer 33(12), 2549 (1992).

89. Hill, M.J., Barker, P.A., Barham, P.J., Puig, C.C., in: Polym. Prepr. (Am. Chem. Soc., Div. Polym. Chem.), ACS Meeting "Recent Advances in Polyolefin Polymers", Washington, 195 (1992).

90. van Ruiten, J., Pijpers, T., in: Polym. Prepr. (Am. Chem. Soc., Div. Polym. Chem.), ACS

Meeting "Recent Advances in Polyolefin Polymers", Washington, 204 (1992).

91. Shirayama, K., Okada, T., Kita, S.-I., J. Polym. Sci.: Part A 3, 907 (1965).

92. Schouterden, P., Groeninckx, G., Van der Heyden, B., Jansen, F., Polymer 28, 2099 (1987).

93. Hosoda, S., Polym. J. 20(5), 383 (1988).

94. Springer, H., Hengse, A., Hinrichsen, G., J. Appl. Polym. Sci. 40, 2173 (1990).

95. Scholte, Th.G., Meijerink, N.L.J., Schoffeleers, H.M., Brands, A.M.G., J. Appl. Polym. Sci. 29, 3763 (1984).

DEVELOPMENT OF HIGH PERFORMANCE TREF

FOR POLYOLEFIN ANALYSIS

L. Wild and C. Blatz

Quantum Chemical Corporation
USI Division
11530 Northlake Drive
Cincinnati, OH 45249

ABSTRACT

Temperature-rising elution fractionation (TREF) has become well established as a crystallizability-based separation technique for polyolefins. The structural information generated by TREF complements the MWD data provided by SEC. Recent advances in TREF technology have lead to the development of analytical TREF systems which are beginning to match SEC in terms of instrumental analytical TREF systems sophistication and quality of separation. The development of such a high performance TREF system is described in this paper with emphasis on improved convenience and capability of operation and enhanced resolution. This instrumental system has the capability of operating at sub-ambient temperatures and is interfaced with a versatile data system which allows a variety of TREF applications to be readily performed. These include the determination of branching distributions in various types of polyethylenes and copolymers and the quantitative analysis of a number of polyolefin blend types including those containing non-crystalline components. Examples are provided to illustrate the effectiveness of the improved analytical TREF.

INTRODUCTION

Temperature rising elution fractionation (TREF) has become widely used for characterizing polyolefin materials in terms of their crystallizability.[1] Recent advances in TREF operations have significantly improved both the convenience of operation and the efficiency of separation to the point where it can be considered for routine analysis in the way that size exclusion chromatography (SEC) is used for MWD determination.[2,3] The development of such a system is outlined here and examples given to illustrate how high performance TREF can be effectively used for the analysis of polyolefin compositions.

HIGH PERFORMANCE ANALYTICAL TREF DEVELOPMENT

The development of analytical TREF (A-TREF) has essentially been one of modifying the existing SEC technique to allow polymer separation as a function of temperature. Thus the same solvent delivery and detector system that was designed for SEC have been used in A-TREF instruments. The constant temperature oven which holds the SEC columns has been replaced with a temperature programmed oven. The major difference between the two techniques is that in TREF a crystallization step, involving slow cooling a polymer solution, is necessary to achieve fractionation in terms of crystallizability. This contrasts the molecular size separation which is produced in SEC by passing a polymer solution through a bed of porous particles.

The need for the controlled crystallization step makes it difficult to accommodate an automatic sample analysis system due to the long time required to effectively complete this pre-fractionation.[4] This problem has been overcome in a recent automatic system which has four TREF units operating simultaneously but in differing stages of the overall fractionation cycle.[3] With this system, it is possible to perform eight fractionations within a 24 hour period even while allowing 10 hours for each cooling step. However, the expense and complexity of operation of such a system makes it unsuitable for most potential TREF users. In our attempts to develop a high performance A-TREF which is both economical and convenient to operate, the following objectives were established: high resolution separation, rapid and convenient operation, accurate and reproducible separation temperature measurement, high signal to noise ratio, and sub-ambient operation for low crystalline materials.

In order to achieve these objectives the following operational procedures have been incorporated into the analytical TREF system:

1. The crystallization step is conducted in an oil bath containing multiple samples of dilute polymer solutions. The bath is slow-cooled down to a temperature well below $0°C$.

2. The crystallized polymer is loaded by co-filtering the polymer/solvent mixture with a coarse filter aid directly into the elution column.

3. The elution step takes place in a temperature-programmed GC oven which is capable of operating from below room temperature. An air oven has the advantage of allowing a rapid rate of temperature rise and a quick turn-around between runs.

4. Use of a matched reference column with an embedded temperature probe is used to allow accurate monitoring of actual separation temperature. This is independent of temperature rise or elution rates even though the column temperature lags the oven temperature.

5. An infra-red detector is used rather than a refractive index detector because of its superior baseline stability in a changing temperature environment. A high detector response along with a high signal-to-noise ratio is achieved by operating at relatively high sample loadings.

6. Data is captured, stored and manipulated by computer to provide a variety of options for expressing the information generated by the A-TREF.

EXPERIMENTAL

A diagram of the basic A-TREF instruments used in these studies is shown in Figure 1. Samples to be analyzed are subjected to the fractionation (crystallization) step in a separate programmed oil bath. The polymer samples are dissolved in hot xylene in capped test tubes

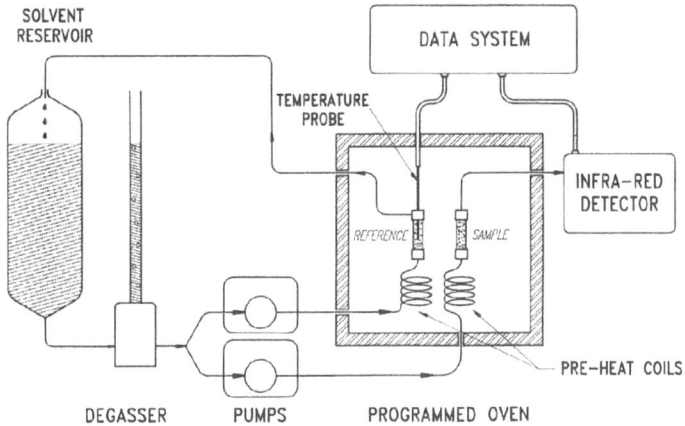

Figure 1. High Performance Analytical TREF System.

which are then held in a metal rack in the bath for slow cooling. A cooling rate of 5°C/hour is normally used and is maintained from 120°C down to a temperature of -20°C. Crystallized samples are then stored in a freezer awaiting analysis. The crystallized polymer is dispersed into a fine precipitate by adding cold acetone and is loaded directly into the elution column by co-filtering with a coarse filter-aid. When samples with low crystallinity components are to be analyzed, this step is performed while maintaining all components below 0°C.

The loaded column is then mounted inside the GC oven which, when necessary, is also maintained at 0°C and solvent elution begun. The solvent used is o-dichlorobenzene which provides a suitable IR window for detecting polyolefins. The temperature program is initiated once a stable baseline is achieved and the detector response recorded as a function of temperature. Current operations use a sample size of 10-15 mgs dissolved in 10 mls of xylene for the crystallization step. Solvent elution rate is maintained at 3mls/min and a rate of temperature rise of 4°C/min. is used.

RESULTS AND DISCUSSION

Effect of Operating Conditions

The development of an analytical TREF was undertaken with the objective of achieving convenient and rapid operation without compromising the desired high resolution and accuracy of the separation temperature. In order to monitor the effects of changes in operating conditions on the effectiveness of separation, a three component test sample has been prepared. The efficiency of separation can be judged by both the accuracy of the separation temperatures and the sharpness of the elution peaks. Efforts have been directed toward reducing the analysis time and improving detector response by increased sample loading. The importance of the dual column system and the slurry loading technique in allowing this to be achieved without loss of resolution is particularly noteworthy.

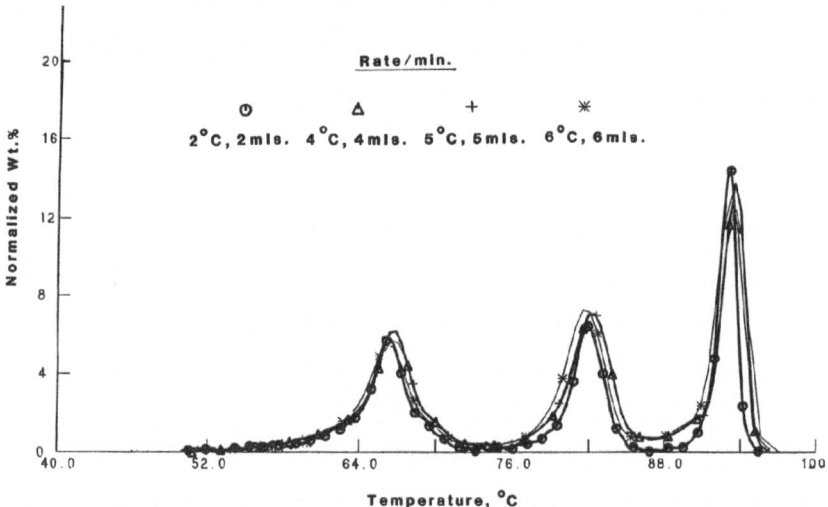

Figure 2. Effect of Rate of Temperature Rise on TREF Separation

Rate of Temperature Rise. With the dual columns in place, a series of fractionations of the three part blend were performed at rates of temperature rise which were increased stepwise from the standard 2°C/min to 6°C/min. The 4 elution rates were also raised in a corresponding manner from 2 mls/min to 6 mls/min to maintain the same overall polymer concentration in the eluent. The resulting curves are shown in Figure 2. It is observed that in spite of the 3-fold decrease in the elution time, baseline resolution continues to be achieved with little variations in peak separation temperatures or peak broadening.

Increase in Loading Level. Obtaining an improved detector response through an increase in sample loading is always desirable if it can be achieved without compromising the fractionation process. In earlier TREF systems there has been a limitation based on the need to maintain the polymer in a narrow band on the column while conducting the crystallization under dilute solution conditions. This implies the use of the lowest possible polymer concentration detectable in order to assure good fractionation. Our present sample preparation technique, in which the crystallization and loading steps are separated, eliminates this constraint. Because of this, it is possible to use increased sample loadings while still conducting the crystallization under dilute solution conditions. This has lead to much improved signal-to-noise ratios through operating the detector at lower sensitivity with no loss of resolution. The results indicate that the higher loading and subsequent improved detector response can be used with no loss of resolution.

Revised Operating Conditions. The operating conditions for the analytical TREF were revised based on the above findings. This resulted in the method of operation described in the experimental section. One change was also made in the data acquisition system to accommodate the higher rates of elution and temperature. This involved increasing the sampling rate for the detector and column temperature to eliminate some truncation of the very sharp peaks which were noted at lower sampling rates. The effectiveness of the revised

150

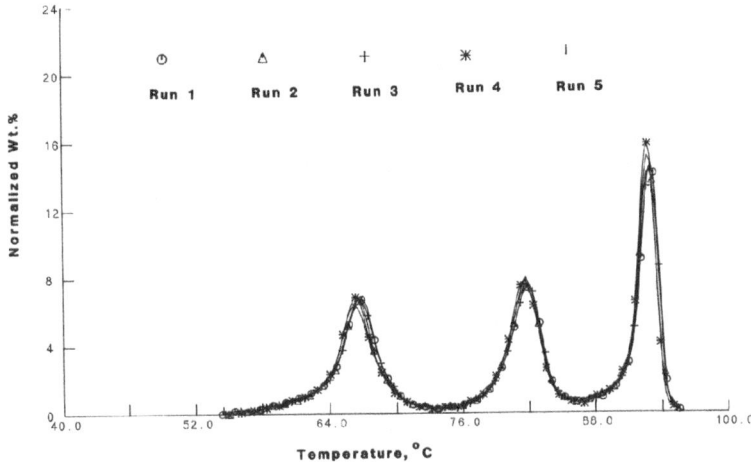

Figure 3. Reproducibility of 3-Part Blend Analysis

conditions can be seen from a series of replicate runs of the 3-part blends shown in Figure 3. The corresponding peak temperatures are listed in Table 1 demonstrating the excellent reproducibility being achieved. Operating under these conditions allows one to analyze up to six samples a day without an auto-injection system. This compares favorably with SEC operating in the auto-inject mode.

Table 1 - Fractionation of 3-Part Blend: Reproducibility Study

Peak Temp, °C

Run #	A	B	C	$\overline{N}$*
1	67.00	81.89	92.76	11.77
2	66.45	81.55	92.75	11.51
3	66.79	81.68	92.44	11.10
4	66.62	81.95	92.92	11.65
5	66.64	81.75	92.74	11.75
6	66.34	81.34	92.56	11.66
7	66.22	81.55	92.54	11.69
	$\overline{X}$=66.58 S.D.=0.27	$\overline{X}$=81.67 S.D.=0.21	$\overline{X}$=92.67 S.D.=0.17	$\overline{X}$=11.59 S.D.=0.23

*$\overline{N}$ = Avg. CH_3/1000C atoms calculated using calibration (Fig. 4)

APPLICATIONS

Comonomer Distributions in Ethylene/-olefin Copolymers

One advantage of well-defined crystallinity separation is that it makes it possible, through application of a calibration curve, to convert the TREF weight distribution into a comonomer content distribution. This has been described earlier[5] but a number of improvements have been incorporated into the TREF operation, a new calibration has been established. To this end a preparative TREF fractionation was undertaken on an ethylene-butene-1- LLDPE resin (1MI, .922 density) and the fractions analyzed by C-13 NMR and subjected to analytical TREF. From this data a calibration curve of ethyls per thousand C atoms vs separation temperature has been constructed (Figure 4). This calibration curve has been successfully applied to the analysis of a variety of ethylene-butene-1 copolymers with excellent agreement being obtained between the measured comonomer content of the whole polymer (by C-13 NMR) and that calculated from the computed comonomer distribution (Figure 5). There are indications that differing comonomers will require a specific calibration to obtain an accurate comonomer distribution due to some effect of branch size on the level of crystallinity reduction.[6] Calibrations for this purpose are being planned.

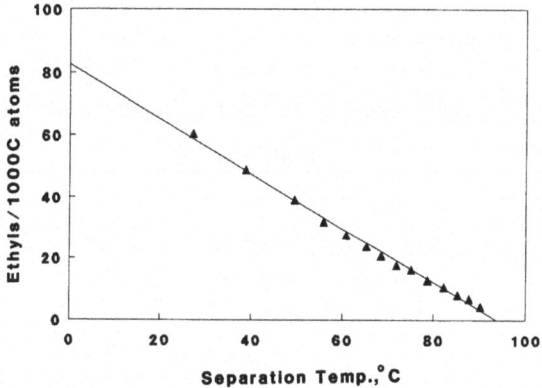

Figure 4. TREF calibration curve established from NMR measurements on ethylene-butene-1 fractions obtained by preparative TREF of a commercial LLDPE.

Analysis of Blends with Crystalline Components

One of the strengths of the TREF method for blend analysis over the commonly used DSC method is that all components are readily detected in a manner which only depends on the polymer's IR extinction coefficient. Thus low crystalline polyolefins are detected with a sensitivity equal to that of the high crystalline components. In a TREF trace in which the components are completely separated, the composition of the blend is measured directly from the ratio of the peak areas.

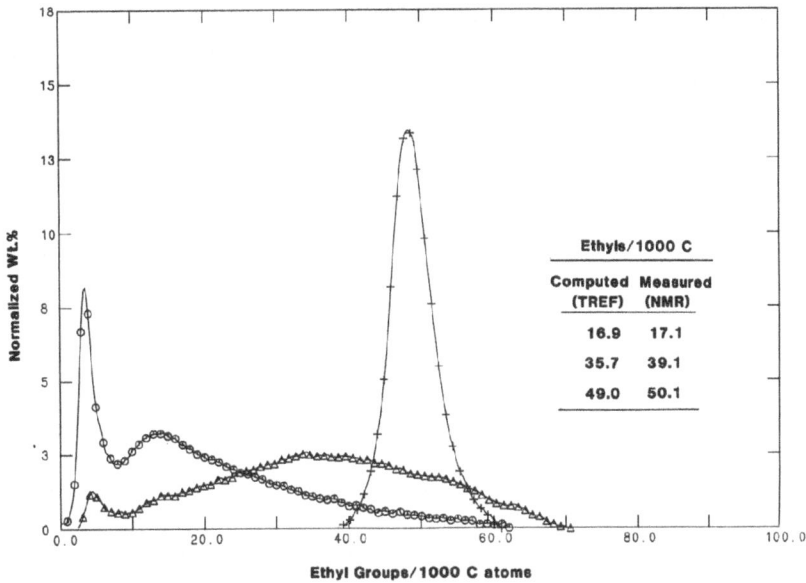

Figure 5. Comparison of Comonomer Distributions for Three Commercial Ethylene-Butene-1 Copolymers.

TREF analysis of the blends where both components are crystalline has been reported earlier.[7,8] As an example of how well the new A-TREF can handle quantitative blend analysis, a series of HDPE/EVA (9wt% VA) blends of known composition have been analyzed. The results obtained are given in Table 2 with the computed values being determined by direct comparison of the peak areas associated with each component. Excellent agreement is obtained as might be expected for blends in which the components are of relatively narrow crystallinity distribution and well separated.

Analysis of Blends Containing a Non-crystalline Component

Up to the present, no reliable way has been devised to quantitatively detect non-crystalline material which may be present in a polyolefin blend. The problem is that although such a material, for example an amorphous EVA or EPR copolymer, is detectable by the IR detector in practice the material is eluted before a baseline can be established. In spite of this, quantitative measurement of the non-crystalline component is still possible by an indirect

Table 2 - Blend Compositions from Peak Areas

Blend Components		% EVA in Blend				
HDPE [1]	Actual	88.6	74.3	50.7	27.1	16.8
EVA [2]	Computed	87.7	74.6	49.2	26.7	17.3

[1] 6 MI, 0.958 density   [2] 5 MI, 9 wt% VA

approach. For example, if a known weight of a blend composition is loaded into a TREF column and the crystalline portion measured quantitatively, then by difference, the amount of the unseen component can be determined. In order to convert peak area into sample weight while accommodating any changes in detector sensitivity, a narrow distribution internal standard has been incorporated. The standard was obtained by a preparative TREF separation, taking the of a relatively narrow LDPE resin. Recovery of the fraction between 61° and 65°C gave a sample consisting of 27.6wt% of the whole polymer with a very narrow short-chain branching distribution (Figure 6).

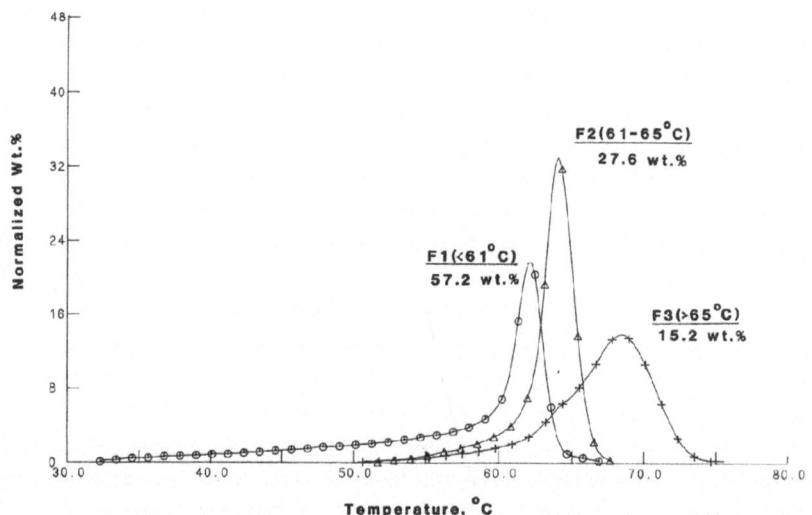

Figure 6. TREF Fractionation of LDPE to Produce a Sharp SCB Distribution Fraction as an Internal Standard.

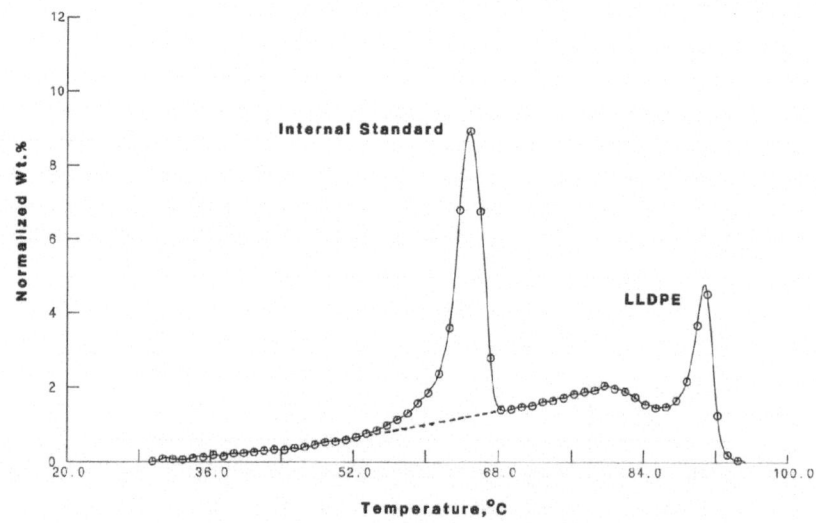

Figure 7. TREF Curve of EVA/LLDPE/Internal Standard Blend.

154

To illustrate how well this approach works in practice a series of HDPE/EVA (35wt% VA) blends were made and analyzed using a known weight of the internal standard. The area/mg of the standard represents 100% and the area/mg of the HDPE peak represents the fraction of HDPE present in the HDPE/EVA blend. The results obtained are given in Table 3. The excellent agreement indicates that the A-TREF using an internal standard can be used for quantitative analysis of blends even if they contain a non-crystalline component. The high resolution capability of the new A-TREF also makes it possible to conduct quantitative analysis where there is some overlap between the crystalline component and the internal standard. Thus for a blend of EVA (35wt% VA) with a LLDPE one obtains the distribution curve shown in Figure 7. The area of the two crystalline components can still be determined by drawing a baseline for the internal standard as shown and allows the blend to be analyzed. This is illustrated by the results obtained for such blend series in Table 3.

Table 3 - Blend Composition Using Internal Standard

Blend Components		% EVA in Blend			
HDPE [1] EVA [2]	Actual	59.0	50.1	40.6	20.1
	Computed	58.8	50.0	42.0	20.0
LLDPE [3] EVA [2]	Acutal	58.4	48.6	40.5	25.7
	Computed	57.7	51.4	39.1	24.1

[1] 6 MI, 0.958 density   [2] 25 MI, 35 wt% VA   [3] 1 MI, 0.923 density

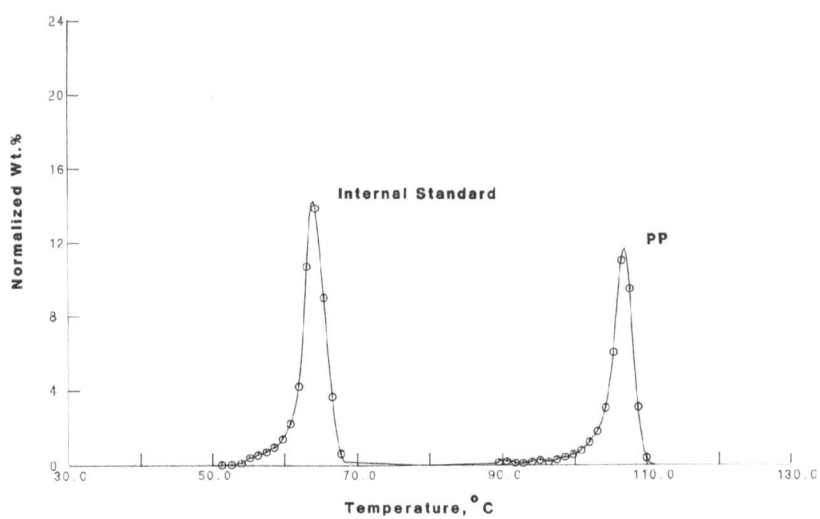

Figure 8. TREF Curve of an Impact PP with Internal Standard.

One of the more useful applications of this technique is in the analysis of impact polypropylenes. This is particularly true for evaluating commercial impact PP products where the PP/EPR blend is produced in situ in a dual reactor process. In this case, the EPR component is not available as an individual component and so the composition must be inferred from analysis of the final product blend. A-TREF analysis of a typical impact PP using the internal standard yields the distribution curve shown in Figure 8 from which one can calculate the weight % of the PP component. To confirm the effectiveness of this technique for impact PP analysis, a series of PP/EPR blends of known composition have been analyzed. The data obtained is given in Table 4 which shows good agreement between the actual and computed values for EPR content.

Table 4 - Polypropylene - EPR Blends by Internal Standard Method

Blend Components		% EPR in Blend			
Polypropylene	Actual	16.0	26.1	36.1	44.9
EPR	Computed	14.2	24.9	36.6	46.8

CONCLUSION

Experimental improvements in the operation of analytical TREF have lead to a system which is capable of quantitative analysis of polyolefins and polyolefin blends. The speed and convenience of operation is comparable to SEC which suggests that a combination of the two complementary techniques will become the approach of choice for routine structural evaluation of polyolefin compositions. As the newer generations of polyolefins emcrge from the single site, constrained geometry catalysts now being commercialized, it is expected that a high performance A-TREF of the kind described here will prove of increasing value for structural evaluation.

Acknowledgements

The authors would like to thank the many colleagues who have contributed to the development of the TREF technique and Quantum, USI Division for support and permission to publish this work.

REFERENCES

1. WILD, L. and KNOBELOCH, D.C. "Process for Fractionating Polymers" US Pat 5,030,713 (1991)

2. WILD, L. "Temperature Rising Elution Fractionation" Advances in Polymer Science 98, pp 1-47, Springer-Verlag (1990)

3. HAZLITT, L.G. "Determination of Short-Chain Branching Distributions of Ethylene Copolymers by Automated Analytical Temperature Rising Elution Fractionation (Auto-ATREF)" J Appl Polym Sci, Appl Polym Symp 45 (BARTH, H.G. ed.) pp 25-37, John Wiley and Sons (1991)

4. WILD, L. and RYLE, T. "Crystallizability Distributions in Polymers: A New Analytical Technique" Polym Prep Am Chem Soc Div Polym Chem 18:182 (1977)

6. HOSADA, S. ''Structural Distribution of Linear Low-Density Polyethylenes'' Polym J 20(5): 383 (1988)

7. KELUSKY, E.C., ELSTON, C.T. and MURRAY, R.E. ''Characterizing Polyethylene-Based Blends with Temperature Rising Elution Fractionation (TREF) Techniques'' Polym Eng Sci 27:1562 (1987)

8. KNOBELOCH, D.C. and WILD, L. ''Analysis of LLDPE/LDPE Blends'' Polyolefins IV Reg Tech Conf Soc Plast Eng, 427 (1984)

NMR ANALYSIS OF MULTICOMPONENT POLYOLEFINS [‡]

H. N. Cheng

Hercules Incorporated, Research Center
500 Hercules Road
Wilmington, Delaware 19808

ABSTRACT

Because many Ziegler-Natta polymers contain multiple components, characterization of these polymers is often difficult. For example, calculation of reactivity ratios through comonomer feed/copolymer composition or through NMR data tends to break down when the copolymer is made up of multiple components. A computer-assisted analytical methodology is described here that enables the reactivity ratios for the separate components to be determined from NMR and comonomer feed ratio data. In the case where NMR and fractionation data on the polymer are both available, a suitable data treatment can provide detailed information on the separate components. The analytical methodologies involved in the analysis are illustrated for the ethylene/propylene and the ethylene/1-hexene copolymers. An ethylene/1-hexene copolymer is found to contain four separate components.

INTRODUCTION

Polymers made with heterogeneous Ziegler-Natta catalysts are known[1-4] to contain multiple catalytic sites, each site potentially having different stereospecificity, regiospecificity, and activity. The polymers resulting from such catalysts are in-situ blends of polymer components which may have different stereochemistry or regiochemistry (for homopolymers) or sequence distribution (for copolymers). The NMR spectra of such polymers are complex and require care to extract the appropriate microstructural information[5-10].

Recently, there have been several reports[10-18] describing systematic analyses of the NMR data of these polymers. A general computerized methodology[10,11] has been proposed. It has been this author's experience that a detailed understanding of the multicomponent behavior of these polymers requires one or more of the following approaches:

[‡] Hercules Research Center Contribution Number 2170

New Advances in Polyolefins, Edited by
T.C. Chung, Plenum Press, New York, 1993

Approach 1. To fractionate the polymer and analyze the fractions by NMR. Through suitable computation, the NMR/fractionation data can be resolved into different polymer components. The components may correspond to the number of active catalytic sites present in the catalyst.

Approach 2. To sample the polymerization vessel at various conversions and again obtain the NMR data of all samples. The NMR/fractionation data can then be analyzed using the mixture analysis[10].

Approach 3 (for copolymer studies). To make copolymers with different compositions using the same catalysts under similar reaction conditions (e.g., temperature, solvent, agitation rate, and reactor geometry), carry out NMR analysis, and develop the computational technique to treat such data.

Approach 1 requires an extra procedure (fractionation), but in suitable cases can indeed provide the requisite information on the multiple components. The approach has been successfully applied to the copolymers of ethylene/propylene[10,11,13,14,18], propylene/1-butene[10], ethylene/1-butene[12], ethylene/1-hexene[16], and ethylene oxide/epichlorohydrin[15] (for sequence distribution), and polypropylene[10,17] and poly(1-butene)[11] (for tacticity). The amount of information available depends on the nature of the fractionation and the quality of the NMR data. Particularly useful is the technique of temperature rising elution fractionation[19,20] (TREF). The combination of TREF and ^{13}C NMR has been shown[14] to provide detailed information about the multicomponent nature of ethylene/propylene copolymers.

Approach 2 permits the kinetics of the copolymerization and the changes in polymer composition and sequence as a function of conversion to be determined. Unfortunately, the NMR data for such kinetic studies are rarely published in the open literature. Nevertheless, the data treatment is identical to *approach 1*; it will not be separately discussed here.

In the published literature on copolymerization, it is customary to keep the catalyst/initiator system constant, and vary the feed ratio of the comonomers. In this way apparent reactivity ratios for the comonomers can be determined (e.g., with Fineman-Ross[21] or Kelen-Tudos plots[22]). The NMR data of the various copolymers made in these studies are sometimes available. If an appropriate analysis is carried out, discrimination can be potentially made as to the presence of single components versus multiple components (*approach 3*).

In this work the analytical methodology for *approach 3* is developed and is applied to published data on ethylene/propylene copolymers and ethylene/1-hexene copolymers. In addition, the NMR/fractionation analysis (*approach 1*) is reviewed and applied to the TREF/NMR data of an ethylene/1-hexene copolymer.

ANALYTICAL METHODOLOGY

The methodology described herein relies on the computerized analytical (model fitting) approach. In this approach[23,24], an approximation of the copolymerization reaction is made via a statistical model, characterized by reaction probabilities. We then associate the intensity of every resolvable NMR line (or alternatively the NMR triad or tetrad sequence) with a theoretical probability expression involving the reaction probabilities. The theoretical and the observed intensities for all the spectral lines are compared and minimized by varying the reaction probabilities systematically. If a satisfactory agreement between the observed and the theoretical intensities is obtained, then the polymer system in question can be described within the context of the data by the statistical model.

160

Program TRIADY (*Approach 3*). When one deals with the NMR data of copolymers (consisting of comonomers A and B) made with different monomer feed concentrations (f_a and f_b), the reaction probabilities do not stay constant among the various copolymer samples. However, the reactivity ratios, r_a and r_b, would indeed be constant for a given polymer component:

$$P_{ab} = \frac{1}{1 + r_a(f_a/f_b)} \tag{1a}$$

$$P_{ba} = \frac{1}{1 + r_b(f_b/f_a)} \tag{1b}$$

where P_{ij} is the first-order Markovian probability of monomer j adding to polymer chain terminating in monomer residue i.

A special case arises when a component obeys the Bernoullian (B) statistics. Thus, $r_a r_b = 1$; $P_{ba} = P_a$, $P_{ab} = P_b$, where P_a and P_b are the Bernoullian probabilities of A and B respectively, and $P_a + P_b = 1$. In this case,

$$P_a = \frac{r_a f_a}{r_a f_a + f_b} , \qquad P_b = \frac{f_b}{r_a f_a + f_b} . \tag{2}$$

In this approach one needs both the NMR sequence data as well as the information on f_a and f_b for all the copolymer samples in order to solve for r_a and r_b.

For a polymer system containing multiple components, a statistical mixture model is needed to describe the polymer. If there are i components in a polymer and w_i and F_i are the respective component weight factors (in mole %) and the probability expressions, then the calculated total intensity of a given line (or sequence) is given by the following expression:

$$I_{total} = \Sigma \, w_i \, F_i \tag{3}$$

For example, the theoretical probability expressions for Bernoullian copolymers containing two Bernoullian components 1 and 2, and three Bernoullian components 1, 2, and 3, are shown in Table 1.

Table 1. Probability expressions[a] for comonomer sequence triads for two-component B/B and three-component B/B/B mixtures

Triad	B/B Mixture Model	B/B/B Mixture Model
AAA	$w_1 P_{a-1}{}^3 + w_2 P_{a-2}{}^3$	$w_1 P_{a-1}{}^3 + w_2 P_{a-2}{}^3 + w_3 P_{a-3}{}^3$
AAB	$2w_1 P_{a-1}{}^2 P_{b-1} + 2w_2 P_{a-2}{}^2 P_{b-2}$	$2w_1 P_{a-1}{}^2 P_{b-1} + 2w_2 P_{a-2}{}^2 P_{b-2} + 2w_3 P_{a-3}{}^2 P_{b-3}$
BAB	$w_1 P_{a-1} P_{b-1}{}^2 + w_2 P_{a-2} P_{b-2}{}^2$	$w_1 P_{a-1} P_{b-1}{}^2 + w_2 P_{a-2} P_{b-2}{}^2 + w_3 P_{a-3} P_{b-3}{}^2$
ABA	$w_1 P_{a-1}{}^2 P_{b-1} + w_2 P_{a-2}{}^2 P_{b-2}$	$w_1 P_{a-1}{}^2 P_{b-1} + w_2 P_{a-2}{}^2 P_{b-2} + w_3 P_{a-3}{}^2 P_{b-3}$
BBA	$2w_1 P_{a-1} P_{b-1}{}^2 + 2w_2 P_{a-2} P_{b-2}{}^2$	$2w_1 P_{a-1} P_{b-1}{}^2 + 2w_2 P_{a-2} P_{b-2}{}^2 + 2w_3 P_{a-3} P_{b-3}{}^2$
BBB	$w_1 P_{b-1}{}^3 + w_2 P_{b-2}{}^3$	$w_1 P_{b-1}{}^3 + w_2 P_{b-2}{}^3 + w_3 P_{b-3}{}^3$

[a] P_{a-i}, P_{b-i} = Bernoullian probabilities of A and B respectively.

It has been found previously[11,12] that model fitting gives the best results when the comonomer triad data of two polymer fractions (or samples) are analyzed simultaneously. The general expressions for a two-component polymer are:

Sample 1

$$I'_{total\text{-}k} = w'_1 F_{1\text{-}k} + w'_2 F_{2\text{-}k} \qquad (4a)$$

Sample 2

$$I''_{total\text{-}k} = w''_1 F_{1\text{-}k} + w''_2 F_{2\text{-}k} \qquad (4b)$$

where the subscript -k corresponds to the kth triad.

Similar expressions can be written for the three-component case.

Sample 1

$$I'_{total\text{-}k} = w'_1 F_{1\text{-}k} + w'_2 F_{2\text{-}k} + w'_3 F_{3\text{-}k} \qquad (5a)$$

Sample 2

$$I''_{total\text{-}k} = w''_1 F_{1\text{-}k} + w''_2 F_{2\text{-}k} + w''_3 F_{3\text{-}k} \qquad (5b)$$

This approach is valid only if the copolymer does not exhibit a substantial amount of conversion heterogeneity.

A computer program (called TRIADY) has been written specifically for this purpose. A schematic diagram is shown in Figure 1. The program[25] accepts as input the comonomer feed concentrations (f_a and f_b) and NMR triad sequences for two related copolymer samples. For the two-component (B/B) model it uses Eqs. 2, 4a, and 4b, and starts the iterations to derive the optimal values of reactivity ratios (r_i) and component weights (w_i). For the three-component (B/B/B) model, it uses Eqs. 2, 5a, and 5b.

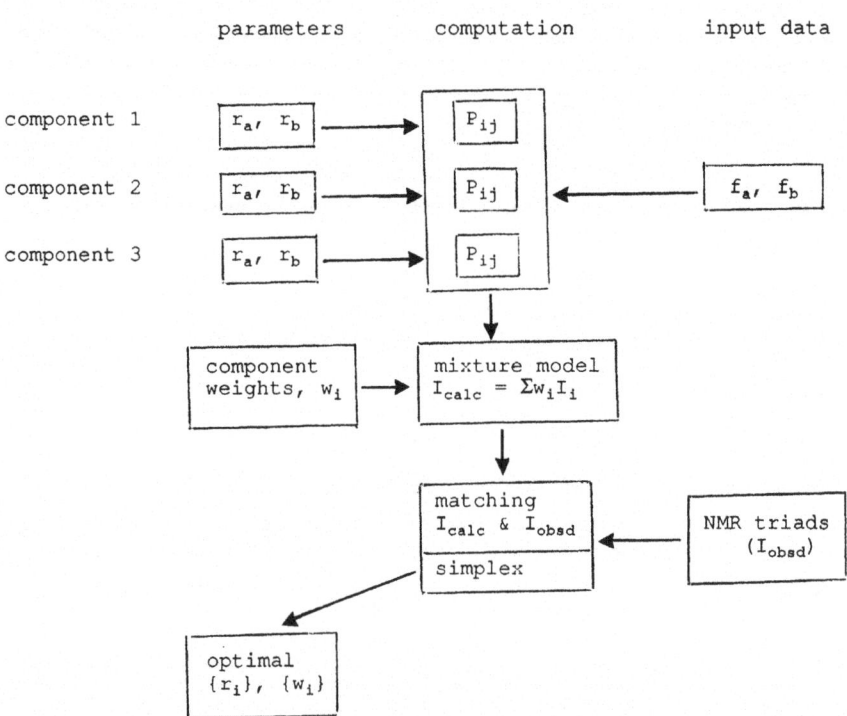

Figure 1. Schematic diagram of program TRIADY.

An alternative computation can be made through a global fitting program (e.g., program MINSQ or SCIENTIST[26]). All the triad data are analyzed by the program simultaneously. An assumed relationship of w_i versus f_a and f_b is taken:

$$w_i = a_i + b_i \left(\frac{f_a}{f_a + f_b} \right) + c_i \left(\frac{f_a}{f_a + f_b} \right)^2 \qquad (6)$$

Optimization can then be carried out to determine the weight parameters (a_i, b_i, c_i) and reactivity ratios (r_a, r_b) for all the components. This global fitting should be done with care to ensure that realistic starting values are used, and that the final values are physically significant.

Program TRIADX (*Approach 1*). The analysis is simplified when one deals with the NMR data of polymer fractions[11,12,14]. Because the fractions arise from the same (starting) polymer, the comonomer feed concentrations should be constant. Thus, the reaction probabilities can be used directly in the two-component equations (Eqs. 4a and 4b) or the three-component equations (Eqs. 5a and 5b) without resorting to the reactivity ratios. As before, the analysis is carried out on the NMR data of two fractions simultaneously. This pairwise analysis is repeated for all consecutive pairs. The end results are the reaction probabilities $\{P_{ij}\}$ and component weights $\{w_i\}$ for all the fractions. Further details on the approach (program TRIADX) have been given elsewhere[11,14,25].

If desired, an additional computation can be made by using a global fitting algorithm such as program MINSQ[26]. In this case, w_i can be parametrized using either the fraction number or the copolymer composition (X), e.g.,

$$w_i = a_i + b_i X + c_i X^2 \qquad (7)$$

For example, X can be the fraction of ethylene in ethylene/1-hexene copolymers.

ETHYLENE/PROPYLENE COPOLYMERS

Soga, et al[27-29] have carried out elegant studies of two types of ethylene/propylene copolymers, one type made with an isospecific catalyst [$TiCl_3$ with $Cp_2Ti(CH_3)_2$] and the other with an aspecific catalyst [$TiCl_3/MgCl_2$ with $Al(C_2H_5)_3$]. A number of samples have been made, and they have determined the feed concentrations of ethylene and propylene (f_E and f_P, respectively). These samples serve as good test cases for *approach 3*.

Aspecific Catalyst: *Approach 3*. Samples A1-A5 were made with an aspecific Ziegler-Natta catalyst[29]. The reported ^{13}C NMR data, the ethylene and the propylene feed concentrations, and the polymer yields are given in Table 2. The analysis is described in detail here to illustrate the use of this approach. The NMR triad data are first fed into the TRIADY program sequentially, two samples at a time. The two-component (B/B) model is first chosen. Computation is fairly straightforward. The component weights (w_i) and the reactivity ratios for propylene (r_P) for both components are obtained. The results are shown in Table 3.

Further computation is carried out by using the results from TRIADY analysis as starting values for a global optimization of the NMR triad data for all five fractions. Again, the two-component (B/B) model is used. The component weight w_1 (from

TRIADY analysis) is first fitted to eq. 6 to give values of a_1, b_1, and c_1. These values are then entered into the global fitting program (MINSQ) together with the NMR triad data. In this way, the optimal values of a_1, b_1, and c_1 and the reactivity ratios, r_P, for the two components are obtained. The results are given in the footnote to Table 3. For comparison, the calculated NMR triad values from this global fitting procedure are given in Table 2. The standard deviation for the fit is 1.5%.

Thus, the ethylene/propylene copolymers contain at least two components. Since the Bernoullian model has been assumed, the reactivity ratios are:

Component 1: $r_P = 0.073$, $r_E = 13.7$.
Component 2: $r_P = 0.503$, $r_E = 2.0$.

It appears then that the catalytic sites responsible for these two components have very different reactivities towards ethylene and propylene. Catalytic site 1 heavily favors ethylene enchainment, whereas site 2 shows only a fourfold preference for ethylene. (It is understood here that the "site" may refer to more than one catalytic active sites that produce the given component. Ideally, of course, each active site corresponds exactly to one component.)

A plot of w_i versus mole fraction of propylene feed (x_P) is shown in Figure 2. From the plot, one can deduce that at $x_P = 0$, $w_1 \approx 0.9$, and at $x_P = 1.0$, $w_1 \approx 0.4$. Thus, in homopolymerization, polyethylene is produced mostly by catalytic site 1, whereas homopolypropylene is made by site 1 (40%) and site 2 (60%).

Soga, et al[29] have also provided the data on the observed polymer yield (y_{total}) for each of the samples. We can determine the polymer yield of each catalytic site (y_i).

$$y_{total} = w_1 y_1 + w_2 y_2 \tag{8}$$

Knowing the values of w_1 and w_2 for all five samples, we can then use linear least squares to determine y_1 and y_2. It turns out that $y_1 = 1.7$, and $y_2 = 5.1$. Thus, catalytic site 2 produces three times as much polymer as site 1.

Table 2. Data on ethylene/propylene copolymers made with an aspecific catalyst[a]. Values in brackets are calculated with two-component (B/B) model; values in parentheses are calculated with three-component (B/B/B) model. Only calculated results from global analysis are shown.

Sample	f_P	f_E	Yield	PPP	PPE	EPE	PEP	EEP	EEE
A-1	0.074	0.0644	2.50	2.8	7.1	7.9	2.8	14.3	65.1
				[1.1]	[4.5]	[8.3]	[2.2]	[16.6]	[67.3]
				(2.0)	(5.5)	(7.6)	(2.8)	(15.2)	(66.9)
A-2	0.145	0.0522	2.69	6.2	10.4	10.4	6.2	22.8	44.0
				[7.0]	[12.7]	[11.1]	[6.4]	[22.3]	[40.4]
				(8.2)	(12.7)	(10.8)	(6.3)	(21.5)	(40.4)
A-3	0.211	0.0408	3.10	18.0	19.7	9.3	9.8	21.3	21.4
				[17.2]	[18.6]	[10.7]	[9.3]	[21.4]	[22.9]
				(16.5)	(18.8)	(10.7)	(9.4)	(21.5)	(23.0)
A-4	0.281	0.0285	3.47	29.6	22.2	9.3	11.1	18.5	9.3
				[31.9]	[21.8]	[8.4]	[10.9]	[16.8]	[10.3]
				(29.5)	(23.5)	(8.2)	(11.7)	(16.4)	(10.6)
A-5	0.359	0.0153	3.50	55.6	18.7	4.5	11.4	6.8	3.0
				[53.8]	[20.7]	[4.3]	[10.4]	[8.5]	[2.3]
				(52.1)	(22.1)	(3.9)	(11.0)	(7.9)	(2.9)

[a] Catalyst is $TiCl_3/MgCl_2-Al(C_2H_5)_3$. NMR data taken from ref. 29.

Table 3. Analysis of ethylene/propylene copolymers made with an aspecific catalyst through pairwise triad analysis (TRIADY) and global analysis (MINSQ)

Sample	2-component model[a]		3-component model[b]		
	w_1	w_2	w_1	w_2	w_3
A-1	0.82	0.18	0.43	0.36	0.21
A-2	0.78	0.22	0.17	0.59	0.24
A-3	0.68	0.32	0.08	0.61	0.31
A-4	0.62	0.38	0.03	0.59	0.38
A-5	0.52	0.48	0.00	0.52	0.48
Reactivity ratio, r_p					
TRIADY	0.08	0.70	0.03	0.11	0.72
MINSQ	0.073	0.503	0.022	0.145	0.847

[a] Global analysis (program MINSQ), for the two-component case, gives:
$$w_1 = 0.896 + 0.116 \, x_p - 0.594 \, x_p^2.$$
$$w_2 = 1 - w_1.$$
[b] Global analysis (program MINSQ), for the three-component case, gives:
$$w_1 = 0.995 - 1.309 \, x_p + 0.364 \, x_p^2,$$
$$w_2 = 0.005 + 0.931 \, x_p - 0.217 \, x_p^2,$$
$$w_3 = 1 - w_1 - w_2.$$
where $x_p = f_p/(f_p + f_E)$.

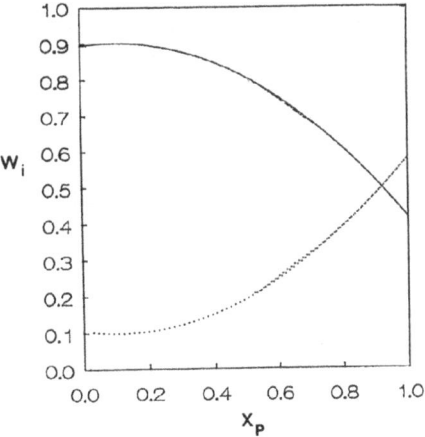

Figure 2. Dependence of w_i on x_p for the ethylene/propylene copolymer made with aspecific catalyst.

It may be noted that, as a rule, if a set of NMR data can be fitted to a n-component model, it can also be fitted to a (n+1) components. We expect, therefore, to be able to fit the data given in Table 2 to a three-component (B/B/B) model. The same procedure is used as before. The NMR triad data are first analyzed by the program TRIADY, and the component weights (w_i) and the reactivity ratios (r_p) for each component obtained (Table 3). These values are then used as starting values for global optimization of data with the program MINSQ. The standard deviation for this three-component model is only slightly lower (1.51%). The calculated triad values are also included in Table 2.

Again, the polymer yield data can be fitted to the three-component model.

$$y_{total} = w_1\, y_1 + w_2\, y_2 + w_3\, y_3 \qquad (9)$$

Using eq. 9 and the yield data in Table 2, we can readily obtain $y_1 = 1.1$, $y_2 = 1.1$, and $y_3 = 8.0$. The choice of the two-component versus the three-component model cannot be made (in this case) from the NMR and polymer yield data alone. This is a limitation often encountered when only one technique (like NMR) is used for investigation. If more definitive information is needed on the number of components present, then additional techniques (e.g., fractionation, TREF, and SEC) are needed. Certainly, if the polymers in question here are fractionated and the fractions analyzed by NMR, then it is possible (in optimal cases) to resolve not only three but even four separate components from the NMR/fractionation data (*vide infra*).

Isospecific Catalyst: *Approach 3.* A similar analysis can be carried out for isotactic ethylene/propylene copolymers. Twelve samples (designated with a prefix B herein) have been published by Soga et al[28], and ten of the data are summarized in Table 4. Samples B-1 and B-12 are homopolymers and are not included in the NMR analysis.

As before, the information on feed composition and triad sequences is first analyzed by program TRIADY. Two samples at a time are analyzed via the two-component B/B model. The component weights (w_i) thus obtained are then fitted to Eq. 6 to determine a_1, b_1, and c_1. These then serve as input to the global fitting program MINSQ where the entire NMR triad data for ten samples are fitted at once. This procedure gives a satisfactory fit to the NMR data (standard deviation = 1.7%). The optimal values for a_1, b_1, c_1 and reactivity ratios are given in Table 4, footnote a. The dependence of w_1 on the mole fraction of propylene feed (x_p) suggests that at $x_p = 0$, $w_1 \approx 0.5$, and at $x_p = 1.0$, $w_1 \approx 0.1$. Thus, in this two-component model homopolypropylene is produced mostly by catalytic site 2, whereas most of homopolyethylene is made almost equally by catalytic sites 1 and 2.

From the observed polymer yields (y_{total}) for the samples, we can also determine the polymer yield for each catalytic site (y_i) through Eq. 8. Sample B-1 and B-12 have unusually lower polymer yields. The enhancement effect of a comonomer on polymer yield is sometimes observed[1]. Soga, et al[28] studied this lower yields and concluded that the effect is purely physical, i.e., due to the higher crystallinity of the homopolymers. If we ignore samples B-1 and B-12, we can readily fit the polymer yield via Eq. 8. The linear least square analysis indicates $y_1 = 1.8$, and $y_2 = 2.0$.

The analysis can be repeated for the three-component (B/B/B) model. In this case, the addition of a third component does not appear to improve the fit. An optimal set of parameters can still be obtained (Table 4, footnote b). The polymer yield data can be similarly fitted to the three-component (B/B/B) model using the values of w_i from the global analysis. Again, samples B-1 and B-12 are ignored. The analysis (Eq. 9) indicates that $y_1 = 1.8$, $y_2 = 3.2$, and $y_3 = 1.0$.

Table 4. Data on ethylene/propylene copolymers made with an isospecific catalyst[28]. Values in brackets are calculated with two-component (B/B) model[a]; values in parentheses are calculated with three-component (B/B/B) model[b].

Sample	f_P	f_E	Yield	PPP	PPE	EPE	PEP	EEP	EEE
B-2	0.0067	0.0766	1.815	0.0	0.1	0.7	0.0	1.7	97.5
				[0.0]	[0.1]	[1.3]	[0.0]	[2.6]	[95.9]
				(0.0)	(0.1)	(1.2)	(0.0)	(2.5)	(96.2)
B-3	0.0134	0.0751	1.820	0.0	0.2	1.5	0.2	3.1	95.0
				[0.0]	[0.3]	[2.2]	[0.1]	[4.3]	[93.1]
				(0.0)	(0.2)	(2.0)	(0.1)	(4.1)	(93.5)
B-4	0.0260	0.0729	1.821	0.0	0.3	2.9	0.6	6.2	90.0
				[0.0]	[0.7]	[3.2]	[0.4]	[6.5]	[89.2]
				(0.0)	(0.6)	(3.1)	(0.3)	(6.3)	(89.6)
B-5	0.0416	0.0712	1.843	0.4	0.9	5.0	0.5	10.5	82.9
				[0.2]	[1.5]	[4.3]	[0.7]	[8.7]	[84.7]
				(0.2)	(1.4)	(4.3)	(0.7)	(8.6)	(84.9)
B-6	0.0528	0.0683	1.906	0.4	1.9	5.6	1.5	11.4	79.3
				[0.3]	[2.2]	[5.1]	[1.1]	[10.2]	[81.1]
				(0.3)	(2.1)	(5.1)	(1.1)	(10.2)	(81.2)
B-7	0.0759	0.0643	2.056	0.8	3.8	6.8	2.5	13.3	72.8
				[0.9]	[4.2]	[6.6]	[2.1]	[13.2]	[73.0]
				(1.0)	(4.2)	(6.6)	(2.1)	(13.3)	(72.8)
B-8	0.153	0.0511	2.081	6.3	10.9	9.8	7.3	18.3	47.4
				[7.5]	[13.5]	[9.0]	[6.8]	[18.1]	[45.1]
				(7.7)	(13.5)	(9.0)	(6.8)	(18.1)	(44.9)
B-9	0.230	0.0379	2.100	16.6	19.5	9.3	11.1	16.9	26.4
				[21.9]	[20.1]	[7.9]	[10.0]	[15.8]	[24.3]
				(22.3)	(19.9)	(7.9)	(9.9)	(15.8)	(24.3)
B-10	0.303	0.0253	2.159	49.2	18.0	5.0	10.3	8.8	8.6
				[41.6]	[20.4]	[5.6]	[10.2]	[11.1]	[11.2]
				(41.7)	(20.2)	(5.6)	(10.1)	(11.3)	(11.2)
B-11	0.401	0.0085	2.208	73.0	13.9	1.7	7.8	2.4	1.2
				[74.9]	[12.2]	[1.8]	[6.1]	[3.7]	[1.2]
				(74.6)	(12.3)	(1.9)	(6.2)	(3.7)	(1.2)

[a] Global analysis (program MINSQ), for the two-component case, gives:

Component 1. $w_1 = 0.537 + 1.139\ x_P - 1.563\ x_P^2$;
$r_P = 0.028$.
Component 2. $w_2 = 1 - w_1$;
$r_P = 0.38$

[b] Global analysis (program MINSQ), for the three-component case, gives:

Component 1. $w_1 = 0.537 - 1.103\ x_P + 1.522\ x_P^2$,
$r_P = 0.028$.
Component 2. $w_2 = 1 - w_1 - w_3$,
$r_P = 0.30$.
Component 3. $w_3 = 0.159 - 0.455\ x_P + 0.875\ x_P^2$,
$r_P = 0.45$.

ETHYLENE/1-HEXENE COPOLYMERS

Approach 3. The ^{13}C NMR assignments for ethylene/1-hexene copolymers have been recently reviewed, and minor corrections made[16]. Soga et al[29] have published copolymerization and ^{13}C NMR triad data for ethylene/1-hexene copolymers prepared with the same aspecific and isospecific catalysts. Pertinent data are summarized in Table 5.

The results of the two-component B/B model-fitting process are shown in Table 5. Consistent values of reactivity ratios are obtained. However, the goodness-of-fit is only fair: the standard deviation in the global fit is 3.1%. A close examination of the observed versus the calculated triad intensities suggests that the deviations are random, perhaps reflecting larger experimental errors. In view of the larger errors associated with this data set, only the two-component B/B model is fitted.

Table 5. Data on ethylene/1-hexene copolymers made with an aspecific catalyst (series C)[a] and isospecific catalyst (series D)[b]. Observed data taken from ref. 29; calculated values in brackets using a two-component (B/B) model

Sample	f_H	f_E	Yield	HHH	HHE	EHE	HEH	EEH	EEE
C-1	0.08	0.0771	2.69	0.0 [0.0]	0.1 [0.4]	4.6 [4.1]	0.0 [0.2]	9.1 [8.3]	86.3 [87.1]
C-2	0.16	0.0771	3.02	0.6 [0.7]	1.2 [2.1]	7.4 [7.3]	2.6 [1.0]	10.3 [14.6]	77.9 [74.3]
C-3	0.40	0.0771	3.67	3.9 [4.2]	5.0 [7.2]	13.3 [11.9]	6.6 [3.6]	18.3 [23.8]	52.9 [49.2]
C-4	0.80	0.0771	5.03	6.1 [9.5]	12.0 [14.3]	15.9 [13.6]	12.1 [7.2]	31.3 [27.2]	22.5 [28.3]
C-5	1.60	0.0771	9.54	26.3 [19.2]	19.4 [22.9]	11.3 [11.5]	13.7 [11.4]	19.4 [23.0]	9.9 [12.0]
C-6	3.20	0.0771	15.8	34.7 [35.4]	32.2 [26.9]	3.9 [6.9]	10.8 [13.4]	18.4 [13.8]	0.0 [3.6]
D-1	0.224	0.0771	1.92	0.0 [0.5]	1.3 [3.5]	3.9 [4.0]	1.8 [2.8]	4.2 [8.1]	88.8 [85.1]
D-2	0.552	0.0771	1.73	2.2 [2.3]	2.5 [1.9]	6.3 [7.9]	3.9 [4.9]	7.4 [6.8]	77.7 [74.3]

[a] Global analysis (program MINSQ), for the two-component case, gives:
Component 2. $w_2 = 0.019 - 0.185 x_H + 0.287 x_H^2$;
$r_H = 0.878$.
Component 1. $w_1 = 1 - w_2$;
$r_H = 0.046$.

[b] Pairwise analysis (program TRIADY), for the two-component case, gives:
Component 1. $w_1 = 0.94$, $r_H = 0.009$.
Component 2. $w_2 = 0.06$, $r_H = 0.363$.
(Samples D-1 and D-2 are only slightly different in w_1 content.)

Table 6. Summary of TRIADY analysis of Soga's polymers

sample series	catalyst	polymer	component 1			component 2		
			r_E	$w_1(X_E=1)$	$w_1(X_E=0)$	r_E	$w_1(X_E=1)$	$w_1(X_E=0)$
A	aspecific	EP	14	0.90	0.42	2.0	0.10	0.58
B	isospecific	EP	36	0.54	0.11	2.6	0.46	0.89
C	aspecific	EH	22	0.98	0.88	1.1	0.02	0.12
D	isospecific	EH	(110)[a]	–	–	(2.7)[a]	–	–

[a] Larger errors are associated with these numbers; see text.

Using this two-component B/B model, we obtained the following results for the aspecific catalyst (Samples C1-C6):

Component 1: $r_H = 0.045$, $r_E = 22.2$.
Component 2: $r_H = 0.88$, $r_E = 1.14$.

Thus, in agreement with the results for the ethylene/propylene copolymers, the aspecific catalyst has at least two catalytic sites, one heavily favoring ethylene enchainment, and the other showing a slight preference for ethylene.

The dependence of w_1 on mole fraction of the 1-hexene feed (x_H) (Table 5, footnote) suggests that at $x_H = 0$, $w_1 \approx 1$, and at $x_H = 1$, $w_1 \approx 0.88$. In homopolymerization, polyethylene is produced entirely by site 1, whereas homopoly(1-hexene) is made by both site 1 (88 %) and by site 2 (12 %). From the polymer yield data for these copolymers, we can determine the yields for the two components via Eq. 8. A least square analysis gives $y_1 = 2.0$, and $y_2 = 18.0$. Thus, site 2 is more productive than site 1. These results are consistent with the results of the ethylene/propylene copolymers.

Soga, et al[29] also provided the data for two ethylene/1-hexene copolymers made with the isospecific catalyst (Table 5). The 1-hexene contents in these samples are low. As a result, only pairwise analysis of the data can be made:

Component 1: $r_H = 0.009$, $r_E = 111$
Component 2: $r_H = 0.363$, $r_E = 2.75$.

Component 1 accounts for approximately 94% of the polymers, whereas component 2 contributes only 6%. Since only two samples are available, the result is only suggestive, and may lack precision.

The analytical results of Soga's polymers are summarized in Table 6. Only the two-component model is used in order to compare the results from both ethylene/propylene and ethylene/1-hexene copolymers. In all four samples, the reactivity ratios (r_E) for the two components are comparable in magnitude. The catalytic site(s) corresponding to component 1 have relatively high activity towards ethylene. The component weights (w_1) are high for the aspecific catalysts (series A and C) but lower in isospecific catalyst (series B). Component 2 is probably a composite of two or more components (*vide infra*); its behavior is just the opposite of component 1.

Approach 1. Kakugo et al[30] have recently prepared an ethylene/1-hexene copolymer using δ-$TiCl_3$ and $Al(C_2H_5)_2Cl$. They carried out a detailed study, using TREF to fractionate the copolymer and analyzing the fractions with ^{13}C NMR. The TREF elution temperature, the weights of the fractions (wt %), and the NMR triad data are given in Table 7. This detailed NMR/fractionation data would be amenable to *approach 1*.

We proceed to use program TRIADX and analyze the NMR data two fractions at a time. After an initial analysis, it appears that fractions 1 and 2 possess a high 1-hexene component; in contrast, fractions 3-8 contain a high ethylene component. further iterations indicate that the polymer contains four components with the following Bernoullian probabilities for 1-hexene (P_H):

component 1: $P_H = 0.89$,

component 2: $P_H = 0.70$,

component 3: $P_H = 0.20\text{-}0.30$,

component 4: $P_H = 0.025\text{-}0.095$.

For comparison, the calculated approximate triad intensities are also given in Table 7. The component weight factors derived from this analysis are shown in Table 8. For better visualization, the component weights are also plotted as a function of copolymer composition in Figure 3. Thus, for this catalytic system there are at least four catalytic sites corresponding to these four components.

Table 7. ^{13}C NMR and TREF data on ethylene/1-hexene copolymers made with δ-TiCl$_3$/Al(C$_2$H$_5$)$_2$Cl. Observed data taken from ref. 30; values in brackets obtained with pairwise fractional analysis (program TRIADX)[a]; values in parentheses obtained with global analysis (program MINSQ)[b]

Fr.	Elution temp (°C)	Ethylene content	wt %	HHH	HHE	EHE	HEH	EEH	EEE
1	− 30	0.28	79.8	50.0	16.0	6.0	8.0	11.0	9.0
				[50.0]	[16.0]	[4.4]	[8.0]	[8.7]	[12.9]
				(51.4)	(15,7)	(4.6)	(7.8)	(9,2)	(11.3)
2	0	0.60	1.5	16.0	13.0	11.0	7.0	20.0	33.0
				[16.0]	[14.0]	[10.0]	[7.0]	[20.0]	[33.0]
				(16.9)	(13.8)	(9.9)	(6.9)	(19.8)	(32.7)
3	30	0.73	1.4	8.0	8.0	11.0	6.0	19.0	48.0
				[8.0]	[10.1]	[9.6]	[5.1]	[19.2]	[48.0]
				(8.2)	(10.1)	(9.2)	(5.0)	(18.4)	(49.0)
4	50	0.78	1.1	6.4	6.4	10.6	4.3	18.1	54.3
				[5.9]	[8.5]	[9.0]	[4.2]	[18.1]	[54.3]
				(5.8)	(8.2)	(8.5)	(4.1)	(17.1)	(56.4)
5	60	0.82	1.4	5.0	5.0	8.0	3.0	15.0	64.0
				[4.5]	[6.4]	[7.4]	[3.2]	[14.7]	[63.9]
				(4.1)	(6.5)	(7.8)	(3.2)	(15.6)	(62.8)
6	70	0.84	2.0	4.0	4.0	7.1	2.0	13.1	69.7
				[3.0]	[4.8]	[6.7]	[2.4]	[13.4]	[69.8]
				(3.4)	(5.5)	(7.4)	(2.8)	(14.7)	(66.2)
7	80	0.92	1.8	2.0	2.0	5.9	1.0	11.8	77.5
				[2.2]	[3.4]	[5.2]	[1.7]	[10.3]	[77.4]
				(1.3)	(1.4)	(5.3)	(0.7)	(10.6)	(80.7)
8	90	0.95	1.9	1.0	1.0	3.0	0.0	6.0	89.0
				[0.5]	[1.0]	[3.0]	[0.5]	[6.0]	[89.0]
				(0.8)	(0)	(4.4)	(0)	(8.7)	(86.5)

[a] Pairwise analysis (program TRIADX) gives the following Bernoullian probabilities: $P_H = 0.89$; 0.70; 0.20-0.30; 0.025-0.095.

[b] Global analysis (program MINSQ), for the four-component case, gives:
Component 1. $w_1 = 1.619 - 3.913\,X + 2.362\,X^2$;
Component 2. $w_2 = -0.466 + 2.328\,X - 1.985\,X^2$;
Component 3. $w_3 = -0.657 + 4.344\,X - 3.841\,X^2$;
Component 4. $w_4 = 1 - w_1 - w_2 - w_3$;
where X = ethylene content. Analysis is based on fixed Bernoullian probabilities: $P_H = 0.89$; 0.70; 0.25; 0.05.

The polymer yield for each component can be calculated by multiplying the weight of the fractions (wt %) with the component weight factors (w_i). The calculation is given in Table 8. The products for each component can be summed together to give the polymer yield for that component. It appears that the catalytic site corresponding to component 1 is the most active (56% yield), followed by sites 3 and 4.

Table 8. Results of the pairwise analysis of TREF/NMR data on ethylene/1-hexene copolymer. Component weight factors and polymer yields

Fr.	wt %	component weights				w_i * wt %			
		w_1	w_2	w_3	w_4	1	2	3	4
1	79.8	0.700	0.052	0.247	0	55.9	4.2	19.7	0
2	1.5	0.078	0.294	0.628	0	0.1	0.4	0.9	0
3	1.4	0	0.220	0.464	0.317	0	0.3	0.7	0.4
4	1.1	0	0.154	0.434	0.414	0	0.2	0.5	0.5
5	1.4	0	0.112	0.259	0.631	0	0.2	0.4	0.9
6	2.0	0	0.069	0.197	0.735	0	0.1	0.4	1.5
7	1.8	0	0.055	0.124	0.822	0	0.1	0.2	1.5
8	1.9	0	0.011	0.046	0.943	0	0.0	0.1	1.8
9	5.8	0	0	0	1.000	0	0	0	5.8
10	2.9	0	0	0	1.000	0	0	0	2.9
sum	99.6					56.0	5.5	22.8	15.2

Table 9. ^{13}C NMR and TREF data on degraded fractions 6 and 7[30], and analysis through program TRIADX.

Fr.	Elution temp(°C)		HHH	HHE	EHE	HEH	EEH	EEE	Component weights[a]		
									w_1	w_2	w_3
D-1	0	obsd.	9.0	11.0	14.0	8.0	21.0	37.0			
		calc.	9.0	13.8	11.1	6.9	22.2	37.0	0.23	0.62	0.16
D-3	50	obsd.	2.0	3.0	6.0	1.0	12.0	77.0			
		calc.	1.4	3.0	6.0	1.	11.9	76.2	0.03	0.17	0.80
D-4	60	obsd.	1.0	1.0	5.0	0	11.0	82.0			
		calc.	0.4	1.5	5.1	0.7	10.2	82.0	0.01	0.10	0.89

[a] Bernoullian probabilities P_H are: 0.69 (component 1); 0.26 (component 2); 0.045 (component 3).

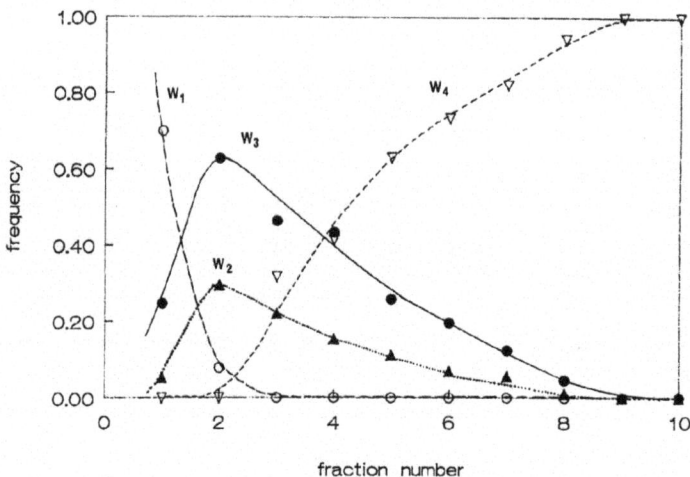

Figure 3. Dependence of w_i on the copolymer ethylene content for the ethylene/1-hexene copolymer.

A further computation can be carried out using global analysis; i.e., taking all the NMR triad data and solving the four-component equation (Eq. 10) simultaneously.

$$I^{\alpha}_{total-k} = w^{\alpha}_1 F_{1-k} + w^{\alpha}_2 F_{2-k} + w^{\alpha}_3 F_{3-k} + w^{\alpha}_4 F_{4-k} \tag{10}$$

where the component weight factors $\{w^{\alpha}_i\}$ are expressed as in Eq. 7, and the superscript α refers to the fraction number. In this analysis, we use the values of $\{w^{\alpha}_i\}$ given in Table 8 as the starting point of the global fit (program MINSQ). The w_1, w_2, w_3, and w_4 values in Table 8 are first fitted individually to Eq. 7 using least square analysis, and the coefficients a_i, b_i, and c_i obtained. The nine coefficients corresponding to the first three components (a_i, b_i, c_i, where i = 1,2,3) are then allowed to vary until the best fits are obtained. The weight factors for component 4 are obtained by difference (w_4 = 1- w_1- w_2 -w_3). In this analysis, the reaction probabilities for the four components are held at P_H = 0.89, P_H = 0.70, P_H = 0.25, P_H = 0.05. Satisfactory fit is obtained in this way (standard deviation = 1.3%).

The calculated triad intensities using the global analysis are also given in Table 7. The optimal coefficients $\{a_i$, b_i, $c_i\}$ obtained are shown in the footnote to Table 7. Using these coefficients, one can readily calculate the component weight factors $\{w^{\alpha}_i\}$. As it turns out, the $\{w_i\}$ values obtained from global analysis (not shown) and pairwise analysis (Table 8) are not substantially different. Thus, the data appear to be well behaved.

Kakugo et al[30] have also taken two fractions (fractions 6 and 7) obtained from the TREF, degraded the polymer, and fractionated it using the TREF. The NMR results for three of the fractions are shown in Table 9. Using the pairwise fractional analysis (program TRIADX), these three fractions (for the degraded polymer) can be readily analyzed. Only three components (components 2, 3, 4) are needed to fit the data; the same P_H values are found to be valid (about 0.70, 0.25, and 0.04). The calculated results (Table 9) indicate that fraction D-1 contains a large amount of component 2. This is not the case in the original fraction 6 and 7 before degradation (Table 7). Thus, the original ethylene/1-hexene copolymer is not only compositionally heterogeneous, but also sequentially heterogeneous.

In their work[30], Kakugo et al proposed that there are four catalytic sites, of which site 2 and site 3 interconvert to form "stereoblock" structures:

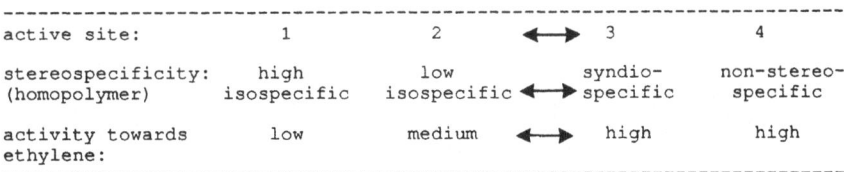

Whereas direct proof of such interconversion is difficult to obtain, the analyses given here are not inconsistent with the proposed mechanism. Indeed, we found four components in the copolymer. The four components do vary in their activity towards ethylene (P_H = 0.89; 0.70; 0.25; 0.05), and there is indeed evidence of sequential heterogeneity in this polymer.

CONCLUSION

This work provides the description of the computer-assisted methodologies used to analyze the NMR data of polymers containing multiple components. Through these means, information on the separate components can be determined. In *approach 3*, the comonomer feed concentrations and the NMR data are analyzed simultaneously to give reactivity ratios and the component weight factors. If the overall polymer yield data are available, the yield of each component can also be extracted.

Approach 1 is geared towards the analysis of polymer fractions. The data treatment provides the reaction probabilities and the component weight factors for each separate component; in addition, the relative polymer yield is available. When used with an efficient fractionation scheme (e.g., TREF), approach 1 can even separate out four components from the NMR data.

REFERENCES

1. P. Pino, P. Cioni, J. Wei, B. Rotzinger, S. Arizzi, in "Transition Metal Catalyzed Polymerizations," edited by R. P. Quirk, Cambridge Univ. Press, Cambridge, 1988, p. 1.
2. Y. V. Kissen, "Isospecific Polymerization of Olefins with Heterogeneous Ziegler-Natta Catalysts," Springer-Verlag, New York, 1985, Chap. 4.
3. P. Corradini, G. Guerra, R. Fusco, V. Barone, Eur. Polym. J., 16, 835 (1980).
4. J. Boor, "Ziegler-Natta Catalysis and Polymerizations," Academic Press, New York, 1979.
5. A. Zambelli, P. Locatelli, A. Provasoli, D. R. Ferro, Macromolecules, 13, 267 (1980).
6. Y. Inoue, Y. Itabashi, R. Chujo, Y. Doi, Polymer, 25, 1640 (1984).
7. J. F. Ross, in "Transition Metal Catalyzed Polymerizations," edited by R. P. Quirk, Cambridge Univ. Press, Cambridge, 1988, p. 799.
8. C. Cozewith, Macromolecules, 20, 1237 (1987).
9. S. Floyd, J. Appl. Polym. Sci., 34, 2559 (1987).
10. H. N. Cheng, J. Appl. Polym. Sci., 35, 1639 (1988).
11. H. N. Cheng, ACS Symp. Ser., 404, 174 (1989).
12. H. N. Cheng, Polym. Bull., 23, 589 (1990).
13. Q. Wu, N.-L. Yang, S. Lin, Makromol. Chem., 23, 589 (1990).
14. H. N. Cheng, M. Kakugo, Macromolecules, 24, 1724 (1991).
15. H. N. Cheng, ACS Symp. Ser., 496, 157 (1991).
16. H. N. Cheng, Polym. Bull., 26, 325 (1991).
17. H. N. Cheng, Makromol. Chem., Theor. Simul., 1, 415 (1992).
18. L. Gilet, M.-F. Grenier-Loustalot, J. Bounoure, Polymer, 33, 4005 (1992).
19. L. Wild, Adv. Polym. Sci., 98, 1 (1990).
20. G. Glockner, J. Appl. Polym. Sci.: Appl. Polym. Symp., 45, 1 (1990).
21. M. Fineman, S. D. Ross, J. Polym. Sci., 5, 259 (1950).
22. T. Kelen, F. Tudos, J. Macromol. Sci., Chem., 9, 1 (1975).
23. H. N. Cheng, J. Chem. Inf. Computer Sci., 27, 8 (1987).
24. H. N. Cheng, J. Appl. Polym. Sci.: Appl. Polym. Symp., 43, 129 (1989).
25. Both programs TRIADX and TRIADY belong to the MIXCO family of programs.
26. Available from MicroMath Scientific Software, Salt Lake City, Utah 84121.
27. K. Soga, H. Yanagihara, D. H. Lee, Makromol. Chem., 190, 37 (1989).
28. K. Soga, H. Yanagihara, D. H. Lee, Makromol. Chem., 190, 995 (1989).
29. K. Soga, T. Uozumi, J. R. Park, Makromol. Chem., 191, 2853 (1990).
30. M. Kakugo, T. Miyatake, K. Mizunuma, Macromolecules, 24, 1469 (1991).

THERMODYNAMICS OF RANDOM COPOLYMER MIXTURES BY SANS

D.J. Lohse, N.P. Balsara, L.J. Fetters, D.N. Schulz, and J.A. Sissano

Corporate Research Laboratories
Exxon Research & Engineering Co.
Annandale, NJ 08801

W. W. Graessley and R. Krishnamoorti

Department of Chemical Engineering
Princeton University
Princeton, NJ 08544

ABSTRACT

A test of the theory of random copolymer miscibility is described in this paper. A series of random ethylene-butene copolymers were made by the saturation of polybutadienes of varying vinyl content, and the interactions in blends of these copolymers were measured by small angle neutron scattering (SANS). It was found that the standard random copolymer theory does not describe the data in that the interaction parameters were not proportional to the square of the difference in the compositions of the component copolymers. Rather, blends of ethylene rich copolymers had lower χ values than corresponding butene rich blends.

INTRODUCTION

Blends of polyolefins are important items of commerce. Despite this commercial importance, very little work has been done to directly determine the thermodynamics of mixing in such blends. There has been some neutron scattering on polyolefin blends, including those of polypropylene and polyethylene[1], those of various forms of polyethylene[2], and those of polypropylene and ethylene-propylene copolymer[3]. The lack of basic data on the thermodynamics of polyolefin mixtures proceeds from two basic causes: the complications due to crystallization for blends in the solid state, and the similarities in physical properties (density, refractive index) between these materials in the melt. Both

New Advances in Polyolefins, Edited by
T.C. Chung, Plenum Press, New York, 1993

factors make it difficult to determine miscibility and interactions at any temperature. This is an unfortunate state of affairs, as one might expect that many polyolefin blends of commercial interest are near phase transitions as they are commonly used. Since there are no strong, specific interactions between these saturated hydrocarbon polymers, they will not be completely miscible over all ranges of temperature, composition, molecular weight, etc. On the other hand, since the dispersive forces between them are so weak (very small unfavorable enthalpy of mixing), it should be possible to find regions of composition and temperature where the blends of high molecular weight polyolefins are single phase, that is, where even the small entropy of mixing is enough to offset the enthalpy of mixing. Developing such an understanding of the miscibility of the polyolefins should have a number of technological implications in terms of the processability and properties of these materials.

From a scientific point of view, polyolefin blends are also of interest as models for polymer blends in general. The polymer mixtures with the most easily understood interactions are those between chemically identical polymers which differ only in their isotopic labelling such as those studied by Bates and coworkers[4]. They measured the interactions between deuterated and hydsrogeneous versions of polystyrene, which can be explained by slight changes in polarizability and density due to the deuteration. Mixtures of structurally different species of saturated hydrocarbon polymers, such as polyolefins, represent the next simplest sort of blends for study due to theior chemical simplicity. Local packing effects, probably negligible in isotopic mixtures, might now become a significant factor because the chain units have different sizes and shapes. Equation of state effects should begin to play some role as well. However, because the polymers are aliphatic hydrocarbons (empirical formula CH_2 in all cases) the change in intermolecular energy with mixing should be relatively small and purely dispersive in origin. Permanent dipole effects and specific interactions are absent. So a more complete understanding of the origins of miscibility in polyolefins should help the understanding of polymer mixtures in general.

The focus of research on the miscibility of polymers is usually the determination of the Flory interaction parameter, χ. This parameter comes from the Flory-Huggins-Staverman[5-7] (FHS) expression for the free energy of mixing two polymers:

$$\frac{\Delta G_m}{VRT} = \frac{\phi_1}{v_1 N_1} \ln \phi_1 + \frac{\phi_2}{v_2 N_2} \ln \phi_2 + \phi_1 \phi_2 \frac{\chi}{v} \tag{1}$$

Here V is the total volume of the sample, R is the gas constant, T is the absolute temperature, N_i is the degree of polymerization of component i (=1 or 2), ϕ_i is the volume fraction of that component, v_i is its molar volume of its mers, and v is an arbitrary reference volume (we use $\sqrt{v_1 v_2}$). Since most polymers of commercial interest have degrees of polymerization of 1000 or more, the first two terms representing the entropy of mixing are generally quite small. Thus the miscibility of the two polymers is largely determined by the value of χ.

One particular question that is of importance in the understanding of polymer blends is that of the mutual miscibility of random copolymers. Many commercial copolymer products have a distribution of composition from chain to chain, due to the polymerization conditions, and so are blends of several copolymers with different amounts of each monomer. This is not necessarily a detriment to the properties of the material and in fact is believed to be an advantage in some cases[8]. In any case it will be important to know the degree of miscibility of the copolymer components with each other. This area has been one of intense theoretical interest. Based on the FHS formulation, Scott[9] and others[10-13] derived the following formula for χ_{12}, the interaction parameter for a blend of two copolymers comprised of monomers A and B with volume fractions of A y_1 and y_2, based on the interaction between the two homopolymers, χ_{AB}:

$$\chi_{12} = \chi_{AB} (y_1 - y_2)^2 \qquad (2)$$

This model assumes that a single parameter representing the interaction between an A repeat unit and a B unit can be used to describe the mixing of any two copolymers, that is, that only very local forces are important and not any long range effects. This has been called the "random copolymer theory." Thus, measurements of χ on several copolymer blends should give the same values of χ_{AB} when treated using Equation 2, and one should be able to predict the values for any copolymer pair once it has been measured on another pair.

In the work described herein, this theory is tested by using small angle neytron scattering to measure the interactions between pairs of ethylene-butene (EB) copolymers over a wide range of composition. Measurements of χ for several copolymers have been performed in the case of styrene-acrylonitrile (SAN) copolymers, but only over a limited composition range[14]. There have also been many copolymers blend systems which have exhibited the so-called "miscibility windows" first described by Karasz et al.[11] and extensively listed in the book of Coleman, Graf, and Painter[15]. However, the interaction parameters have not been directly measured in those cases. Thus, this study not only provides some important information for the EB system, but is also the first extensive and direct test of Equation 2 and the random copolymer theory.

EXPERIMENTAL

Polymers and Characterization

The details of the synthesis and characterization of these model polymers have been given previously[16]. The model ethylene-butene copolymers (EBs) were made by the saturation of polybutadienes of varying vinyl content. The polybutadienes were made anionically to produce polymers with narrow molecular weight distributions and well defined chemical compositions. Each 1,2 addition in the butadiene polymerization translated into a butene moiety in the final hydrogenated polymer and each 1,4 addition into two ethylenes. The level of vinyl incorporation was increased by raising the polarity of the polymerization medium. The saturation was performed with heterogeneous palladium catalysts which provide essentially complete saturation without degrading the chains. As an added advantage, the saturation was done with deuterium as well as with hydrogen, allowing for the production of both labeled and unlabeled versions of each model polyolefin for SANS, our primary means of obtaining the desired thermodynamic information.

Molecular weights and their distributions were determined by light scattering and GPC. Density was measured on a density gradient column; a comparison of the densities of the H and D versions of a given polymer was used to find the level of deuteration in each material. Chemical microstructures were determined by ^{1}H and ^{13}C NMR. All of these polymers were found to have sequence distributions characteristic of random copolymers in the ^{13}C-NMR results. The range of model EB compositions (here designated as HPBxx for xx wt% butene) used in this study is shown in Table I, which also lists the characterization data of these polymers.

Table 1. Molecular Characteristics of Random EBs

Sample	Weight % Butene	M_w (x10³)[a]	% deuteration[b]
HPB38	38	98.9	37.9
HPB52	52	81.8	34.4
HPB66	66	114.5	40.6
HPB78	78	72.2	29.6
HPB88	88	90.4	37.0
HPB97	97	90.0	34.9

a. by GPC; for all HPBs, $1.05 < M_w/M_n < 1.10$.
b. by density measurements.

Neutron Scattering and Data Analysis

All of the small angle neutron scattering measurements were performed at the Cold Neutron Research Facility at the National Institute of Standards and Technology in Gaithersburg, MD. Most of this work was done on the 8 meter apparatus, with some data taken on the 30 meter device. All of the experimental details have been described in the earlier report[16], and only a few will be discussed here. We do show how we work from the interaction parameters (χ's) for the blends of hydrogenous and deuterated polymers to those for purely hydrogenous blends.

The scattering data were analyzed within the general framework of the random phase approximation (RPA), which provides a relationship[17] for the structure factor of a two component, single-phase mixture obeying FHS theory. Although strictly speaking the components in our case are statistical copolymers, we treat them as homopolymers for purposes of analysis. Thus, from RPA and Eq. 1,

$$\frac{1}{S(q)} = \frac{1}{v_1 N_1 \phi_1 P_1(q)} + \frac{1}{v_2 N_2 \phi_2 P_2(q)} - 2\frac{\chi}{v} \qquad (3)$$

where $P_1(q)$ and $P_2(q)$ are the normalized ($P_i(0) = 1$) form factors (partial structure factors) of the component polymers and $S(q)$ is the static structure factor measured for the blend. For the form factors we use the Debye formula for monodisperse random coil polymers. Following the analysis procedures outlined in reference 16, we can thus obtain values of χ for any single phase blend at a variety of temperatures and compositions.

In particular, it is possible to measure both the interaction parameter of a blend where one component (A) is deuterated and the other (B) is not, $\chi_{DA/HB}$, and also that of the "swap", where B is deuterated and A is not, $\chi_{HA/DB}$. In general, these do not have the same value. In a previous paper[18] we have shown how the interaction parameters from the DA/HB and HA/DB blends can be used to find $\chi_{HA/HB}$, the interaction parameter for the completely unlabeled blend of HA and HB. This is possible because the effect of deuteration is generally to reduce the cohesive energy density and so the solubility parameter of a polymer. The interaction parameter between component 1 and 2, χ_{12}, can be related to the difference in the solubility parameters of the blend components, δ_1 and δ_2:

$$\chi_{12} = \frac{v}{RT}(\delta_1 - \delta_2)^2 \qquad (4)$$

where R is the gas constant and T the absolute temperature. Suppose HA has a larger solubility parameter than HB. Then substituting DA for HA will decrease the difference in solubility parameters and so the interaction parameter ($\chi_{HA/HB} > \chi_{DA/DB}$), but substituting DB for HB will increase the difference ($\chi_{HA/HB} < \chi_{HA/DB}$). So $\chi_{HA/HB}$ is intermediate between the χ's for the two labeled blends, and in fact can be well approximated by the average of the two. Thus having labeled versions of each component allows us to derive values for the χ's of unlabeled blends, which are of course the values of technological interest.

RESULTS

We have measured the values of the interaction parameter for five different blends of the random EB copolymers listed in Table 1. These were those with a difference in butene content of between 9 and 14 wt %. Those blends for which there was a greater difference in composition were immiscible at all accessible temperatures. The measured values of χ_{HH} (determined as described above) for all of these blends at a range of temperatures and at a 50/50 composition are given in Table 2.

Table 2. χ_{HH} (x 10^{-4}) of Random EBs at 50/50 Composition

Components	51°C	83°C	121°C	167°C
38/52	11.8	9.5	7.7	6.3
52/66	imm	15.5	12.9	10.3
66/78	imm	12.4	10.7	9.1
78/88	imm	14.1	11.8	9.8
88/97	11.7	10.5	8.7	6.7

imm = immiscible at this temperature and composition.

These results can be used to test the random copolymer theory. Thus, if this model (Equation 3) held for our data, dividing the measured χ by the square of the difference in the butene contents of the components, Δ (= (y_1 - y_2)), would give the same value from each blend. However, as the plot in Fig. 1 shows, this is not the case. The values increase with the butene content. That is, those blends with a lower average value of butene content produce lower values of χ_{AB}, the calculated value of the interaction parameter between the polyethylene and polybutene homopolymers. In other words, EB blends with compositions near PE are "more miscible" than those near PB. So attempting to describe the interactions in EB blends on the basis of a single parameter for the interaction between an ethylene and a butene unit does not work. We can speculate that this discrepancy between the standard random copolymer theory and these EB results is due to some geometrical or packing considerations[19], but there is no theoretical basis for this.

We can also use these results to provide the solubility parameters for the EB copolymers. This can be done by using Equation 4 to derive values for $(\delta_1 - \delta_2)$ from χ_{12}. Since this just gives the difference in the two solubility parameters, two factors are needed to establish values of δ for the copolymers: a general protocol for determining which component has the larger δ in each blend, and the absolute value for one of the copolymers from other data. For the latter, we choose a value for PE, which is the polyolefin for which the best data exists[20], at 167°C based on solubility data and group contribution calculations (17.0 MPa$^{1/2}$). On the other point, we know that the more highly branched copolymer has the lower solubility parameter[18] for several reasons. First, we always measure higher χ values when the high butene copolymer is labeled than when the low one is. Secondly, the heats of vaporization for low molecular weight, branched materials are lower than those of their linear analogs[20]. Finally, group contribution scheme calculations of such solubility parameters[15] also suggest that the branched polyolefins should have lower δ values than their linear counterparts, i.e., the value for PB should be lower than that for PE.

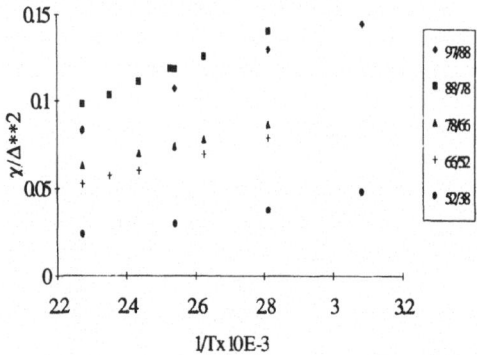

Figure 1. Plot of χ_{HH}/Δ^2 vs. 1/T for blends of random EB copolymers.

In this way a catalogue of δ's for the EB copolymers can be developed. This is shown for 167°C in Table 3 and plotted in Figure 2. We have also included several values of δ from the work of Rhee and Crist[21]. Notice that δ is not a linear function of butene content, which reflects the non-conformance of the χ values to random copolymer theory.

With these values and Equation 4, one can calculate χ for any pair of EB copolymers. Then combining this with the molecular weights of the polymers, one can find the phase diagram of their blend by FHS theory (Equation 1). For polymers with molecular weights of 100,000 (typical of commercial materials), we can express the results by saying that copolymers differing by less than 20 wt% butene will be miscible above 100°C (in the melt) at the high ethylene end, but that this critical compositional difference is about 10% near polybutene.

Table 3. Solubility Parameters of EB Copolymers

Weight % Butene	δ @ 167°C(MPa$^{1/2}$)
0	17.00
7[a]	16.93
22[a]	16.76
27[a]	16.72
35[a]	16.64
38	16.62
48[a]	16.42
52	16.44
66	16.19
78	15.96
88	15.73
97	15.53

a. Derived from data of Rhee and Crist[21].

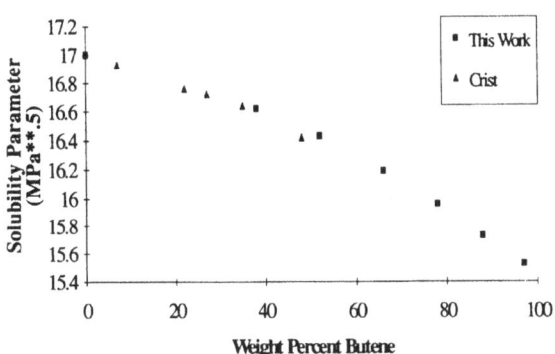

Figure 2. Plot of δ vs. butene content for EB copolymers at 167°C.

CONCLUSIONS

In this study model versions of a wide range of ethylene-butene copolymers have been synthesized and used to determine the interactions in blends of these copolymers by small angle neutron scattering. This is the first direct test of the random copolymer theory in which the χ parameters of copolymer blends were measured on compositions from one homopolymer to the other. It was found that this theory does not adequately describe the data, in that the miscibility does not depend only on the difference in the compositions of the two components. Greater miscibility (smaller χ) was seen for compositions near polyethylene than for those near polybutene. From these data, a catalogue of the temperature and composition dependence of the solubility parameters of EB copolymers has been derived. In the near future we will extend the range of copolymers studied and include results for other model polyolefins as well.

ACKNOWLEDGMENTS

We are grateful to Buckley Crist for discussions and to the National Science Foundation (Grant DMR89-05187) for support of WWG and KR.

REFERENCES

1. Wignall, G. D.; Child, H. R.; Samuels, R. J.; *Polymer* **1982**, *23*, 957.

2. Stehling, F. C.; Wignall, G. D.; *Polym. Prepr. Am. Chem. Soc. Div. Polym. Chem.* **1983**, *24*, 211.

3. Lohse, D. J.; *Polym. Eng. & Sci.* **1986**, *26*, 1500.

4. Bates, F. S.; *J. Appl. Cryst.* **1988**, *21*, 681.

5. Flory, P. J.; *J. Chem. Phys.* **1941**, *9*, 660.

6. Huggins, M. L.; *J. Chem. Phys.* **1941**, *9*, 440.

7. Staverman, A. J.; Van Santen, J. H.; *Recl. Trav. Chim.* **1941**, *60*, 76.

8. Deblieck, R. A. C.; Mathot, V. B. F.; *J. Materials Sci. - Letters* **1988**, *7*, 1276; and Mirabella, F. M.; Westphal, S. P.; Fernando, P. L.; Ford, E. A.; Williams, J. G.; *J. Polym. Sci. - Physics* **1988**, *26*, 1995.

9. Scott, R. L.; *J. Polym. Sci.* **1952**, *9*, 423.

10. Krause, S.; Smith, A. L.; Duden, M. G.; *J. Chem. Phys.* **1965**, *43*, 2144.

11. tenBrinke, G.; Karasz, F. E.; MacKnight, W. J.; *Macromolecules* **1983**, *16*, 1827; Shiomi, T.; Karasz, F. E.; MacKnight, W. J.; *Macromolecules* **1986**, *19*, 2274.

12. Kambour, R. P.; Bendler, J. T.; Bopp, R. C.; *Macromolecules* **1983**, *16*, 753.

13. Paul, D. R.; Barlow, J. W.; *Polymer* **1984**, *25*, 487.

14. Schmitt, B. J.; Kirste, R. G.; Jelenic, J.; *Makromol. Chem.* **1980**, *181*, 1655.

15. Coleman, M. M.; Graf, J. F.; Painter, P. C.; "Specific Interactions and the Miscibility of Polymer Blends", Technomic, Lancaster, Pennsylvania, **1991**.

16. Balsara, N. P.; Fetters, L. J.; Hadjichristidis, N.; Lohse, D. J.; Han, C. C.; Graessley, W. W.; Krishnamoorti, R.; *Macromolecules* **1992**, *25*, 6137.

17. deGennes, P. G.; "Scaling Concepts in Polymer Physics", Cornell University Press, Ithaca, **1979**.

18. Graessley, W. W.; Krishnamoorti, K.; Balsara, N. P.; Fetters, L. J.; Lohse, D. J.; Schulz, D. N.; Sissano, J. A.; *Macromolecules*, in press.

19. Walsh, D. J.; Graessley, W. W.; Datta, S.; Lohse, D. J.; Fetters, L. J.; *Macromolecules* **1992**, *25*, 5236.

20. Barton, A. F. M.; "Handbook of Polymer-Liquid Interaction Parameters and Solubility Parameters", CRC Press, Boca Raton, Florida, **1990**.

21. Rhee, J.; Crist, B.; *Prepr. Am. Chem. Soc. Div. Polym. Mater. Sci. & Eng.* **1992**, *67*, 209.

POLYETHYLENE - COPOLYMER MODEL BLENDS: MORPHOLOGY AND MECHANICAL PROPERTIES

B. Crist and J. Rhee

Dept. of Chemical Engineering
Northwestern University
Evanston, IL 60208

INTRODUCTION

Random copolymers of ethylene and α-olefins made by low pressure polymerization are known to be mixtures of chains which vary considerably in comonomer content.[1] Mirabella, et al.[2] have shown that such composition heterogeneity in linear low density polyethylene (LLDPE, in that case ethylene-butene copolymers) leads to the formation of dispersed, submicron size domains of highly branched polymer, presumably by phase separation in the melt. A similar conclusion was reached by Deblieck and Mathot[3] regarding the morphology of ethylene-octene copolymers. It was further suggested by Mirabella, et al.[2] that the superior plane strain fracture toughness of LLDPE (J_C = 40 kJ/m^2 at -20°C, versus ~2 kJ/m^2 for unbranched polyethylene homopolymer) results from the presence of the dispersed rubber-like regions. Their conclusion was underscored by the observation that conventional low density polyethylene (LDPE) with similar density, branching and molecular weight, but characterized by a narrower composition distribution, has no melt phase separation and a low toughness, about the same as that of unbranched homopolymer (PE).

More germane to the present work are related copolymers, sometimes called medium density polyethylene (MDPE), which have lower average comonomer concentrations and branching distributions which are similar to, but narrower than, those of LLDPE. An example is given by the temperature rising elution fractionation (TREF[4]) curves in Figure 1.[5] Unbranched PE is eluted near 98°C, as indicated by the dashed curve. The copolymer (ethylene-hexene with 4.4 butyl branches/1000 C atoms;[6] National Institute of Standards and Technology SRM 1496) has a bimodal branching distribution. A large fraction of the chains is essentially unbranched, and a measurable quantity is eluted below 60°C, corresponding to more than 15 branches/1000 C atoms. There is no evidence of a second melt phase composed of comonomer-rich chains in such materials[7,8], an observation confirmed by calculation of the equilibrium phase condition for multicomponent random copolymers.[9,10] Such "one phase" systems nevertheless have remarkable resistance to plane strain fracture, as shown by the 10^4 enhancement of constant load lifetimes reported by Brown, et al.[11] for a series of ethylene-hexene copolymers ranging from 0 to 4.6 butyl branches per 1000 backbone carbon atoms. It appears, therefore, that the "second phase" toughening proposed by Mirabella, et al.[2] does not apply to this class of copolymers.

One is thus presented with an intriguing scientific question: by what means does the enchainment of a small amount of "impurity" - less than 0.5% of the backbone atoms are branched - lead to the remarkable increase in fracture toughness of polyethylene? This issue has technological implications as well, since MDPE is becoming the material of choice for many applications; see for instance the *Proceedings* volume cited in reference 7. The number and character of tie molecules in ethylene - α-olefin copolymers have been considered.[12,13] That approach ignores, however, composition heterogeneity which has been shown to be significant in LLDPE[3] and which is present in MDPE. Hosoda and Uemura[14] emphasize the relation between short chain branching distribution and the range of crystal thicknesses formed on cooling. They also consider the effect of size of the short chain branch. Larger branches are more likely to be rejected from crystals, hence are more efficient at reducing crystal thickness, further increasing tie molecule concentration and mechanical strength.

Model binary blends are used in this work to study the relation between composition heterogeneity, morphology and mechanical properties. Blends use unbranched PE and model ethylene-butene copolymers ranging from 15 to 117 ethyl branches per 1000 backbone C atoms. PE and HPB are melt-miscible in two of the blends, and immiscible in the third blend. While these systems do not duplicate commercial MDPE in detail, they are well characterized reference materials which permit definite conclusions to be formulated. Plane strain fracture toughness is shown to be enhanced in systems which are single phase in the melt; the presence of a dispersed rubbery phase has virtually no effect. Tensile properties are also considered.

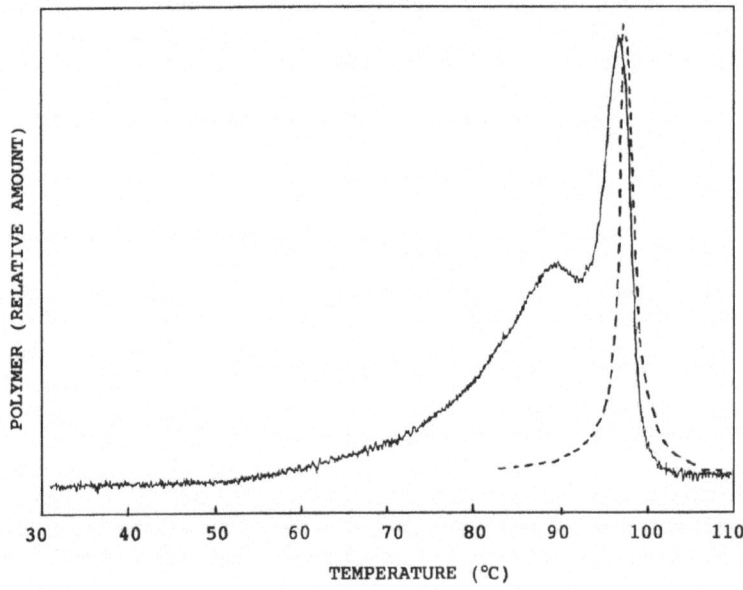

Figure 1. TREF curves of unbranched PE (dashed line) and MDPE ethylene-hexene copolymer. Less branched chains elute at higher temperatures.

EXPERIMENTAL

Polymers

High density polyethylene (PE) was used as the major blend component. To this branch-free polymer was added 5% or 10% by weight of hydrogenated polybutadiene (HPB), a model ethylene-butene random copolymer.[15] HPB synthesis and characterization have been described elsewhere;[15,16] polymer characteristics are summarized in Table I, where N_w is the weight-average degree of polymerization of C_4H_8 monomers and y is the fraction of ethyl-branched C_4H_8 units. Branching is indicated by the suffix $100y$, e.g. HPB06 has 6% branched C_4H_8 repeats. The number of ethyl branches per 1000 backbone C atoms is also given in Table I. Note that HPB has a small polydispersity of chain lengths ($N_w/N_n < 1.2$), and that each HPB is compositionally uniform, i.e. y is the same for all chains in the model copolymer.

Melting temperature T_m, density ρ and volume fraction crystallinity α_v of HPB are also given in Table I. Physical characteristics of these model copolymers have been studied earlier.[15,17] Density and crystallinity of PE depends of course on crystallization conditions, whereas those of HPB are insensitive to cooling rate.

Table I
Polymer Characterization

Polymer	y	br. 1000 C	N_w	N_w/N_n	$T_m[°C]$	$\rho[kg/m^3]$	α_v
PE	0	<0.5	2920	11	135	~970	~0.80
HPB06	0.06	15	2130	--	110	905	0.44
HPB22	0.22	61	1940	--	75	890	0.29
HPB38	0.38	117	2180	1.19	--	860	0

Blends

The three blends of PE and HPB are described in Table II. Components were mixed in xylene at 140°C, precipitated in a 5:1 excess of chilled methanol and dried under vacuum at room temperature. Blend preparation was nontrivial, as some 90 g of each were required for fracture toughness testing. Unblended PE was dissolved and reprecipitated to provide the same history.

Blends 1 and 2 contain the same number of (differently branched) HPB chains, but differ in *average* branch concentration. Blends 2 and 3 have nearly the same average branching, but different concentrations of HPB chains. Most important is the number of melt phases; Blends 1 and 2 are predicted to be single phase, while Blend

Table II
Blend Compositions

Blend	Composition	Avg. br. 1000 C	Melt Phases
1	PE/HPB06(10%)	1.5	one
2	PE/HPB22(10%)	6.1	one
3	PE/HPB38 (5%)	5.9	two

3 is in two phase portion of the phase field map. Figure 2 is based on the effective interaction parameter $\chi = 0.022y^2$ for binary PE/HPB systems at 150°C.[16] Calculation of the "cloud point" curve (solid line) uses the method of Roe and Lu[18] to treat the polydisperse PE component; HPB is assumed to be monodisperse with N = 2100. It should be remembered that the cloud point curve is *not* the binodal, and one cannot infer coexisting compositions or volume fractions from Figure 2. The relevant points are that Blends 1 and 2 are expected to have one phase in the melt, while Blend 3 is will have two phases (in the metastable region bounded by the cloud point and spinodal lines).

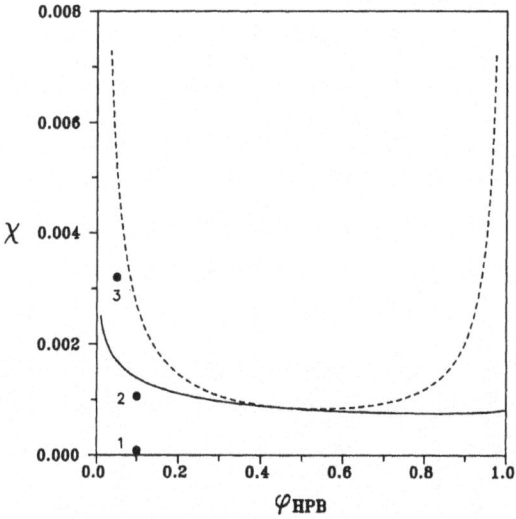

Figure 2. Phase field map at T = 150°C for PE-HPB binary blends with molecular weights used in this study. Ordinate χ changes with HPB branching y, and ϕ_{HPB} is the volume fraction of HPB in the blend. The solid "cloud point" line demarcates one phase (below) and two phase (above) conditions. The dashed line is the spinodal. Points indicate Blends 1 and 2 (one phase melt) and Blend 3 (two phase melt).

Morphology

Dried powder of PE and the three blends was melted at 150°C for times between 20 and 600 sec, then quenched in cold water. The resulting 0.4 mm thick films were ultramicrotomed at -120°C, then etched with an oxidizing reagent which preferentially removes noncrystalline polymer.[16] Since HPB is much less crystalline than PE, the appearance of quenched and etched material indicates the presence or absence of phase separated regions in the melt state. Etched samples were coated with gold and examined by scanning electron microscopy (SEM).

The morphology of thick compression molded samples for fracture toughness testing (see below) was examined in a similar fashion. Of primary interest are the semicrystalline microstructure formed by slow cooling and the evolution of damage zones in notched samples subjected to plane strain. Morphology was observed by SEM of ultramicrotomed surfaces (usually etched).

J Integral Analysis

Plane strain fracture toughness, based on nonlinear elastic fracture mechanics, was measured by procedures established earlier.[19] Plaques 12 mm thick were compression molded from powder using vacuum methods to prevent void formation. While this process involved holding the polymer for about 150 min at temperatures up to 200°C, no oxidation (IR spectroscopy) or molecular weight change (GPC[20]) was detected. Press platens were water cooled, leading to crystallization at a moderate but unknown rate. The molded plaques were machined into deeply notched compact tension (CT) specimens of thickness B = 10 mm and width W = 22.5 mm. The notch of length a_0 = 0.5W was made by a razor cut at the end of a machined cutout. Samples had planar surfaces with no side grooves.

The multiple-specimen regression method was used to establish J_C, the critical value of the J integral required for a crack to initiate at the notch tip. CT samples were deformed at 0.5 mm/min at room temperature, then unloaded and examined for stable crack growth Δa = a-a_0. J was calculated from:[19,21]

$$J = \frac{2\eta U}{B(W-a_0)} \tag{1}$$

Here U is the area under the load-displacement curve and η = 1.13 accounts for stress concentrations caused by the notch when a_0/W = 0.5. Crack growth Δa was measured at the midthickness plane by SEM on transverse sections which had been ultramicrotomed. A plot is made of J *versus* Δa; extrapolation to $\Delta a \approx 0$ establishes J_C for crack initiation. In these materials one always observes a macrocraze at the notch tip before failure. True crack growth occurs by breakdown of oriented material within the macrocraze at $J \geq J_C$.

Tensile Testing

Tensile samples (gauge length 19.5 mm, width 4.0 mm) were cut from 0.5 mm thick films which had been compression molded and cooled at -1°C/min. Tensile modulus and yield stress were measured at room temperature at a strain rate of 4.0×10^{-4}/sec.

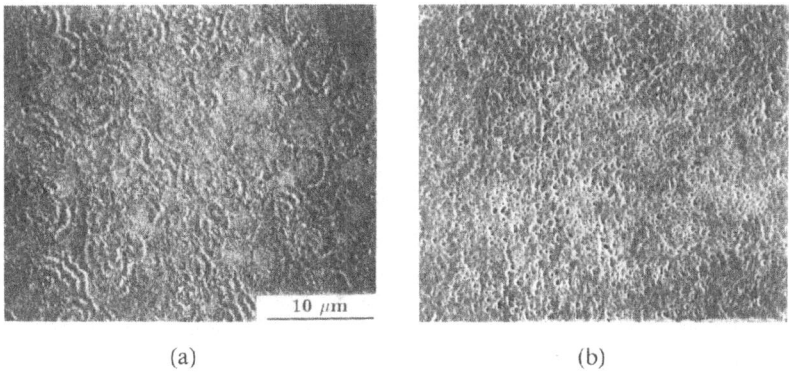

(a)   (b)

Figure 3. Microtomed and etched surfaces from within quenched films. (a) Blend 1; (b) Blend 3 (at same magnification).

RESULTS AND DISCUSSION

Morphology

Quench crystallized thin films were used to establish the validity of the phase behavior predicted by the thermodynamic analysis (Figure 2). Blends 1 and 2 form banded spherulites of ~5 μm diameter which are indistinguishable from those in PE. And example is shown in Figure 3a. We conclude that the melt was single phase and that any segregation of HPB which may occur on rapid crystallization is limited to distances less than spherulite dimensions. Blend 3 is has a quite different appearance caused by two phases in the melt. Figure 3b is for a film made from powder heated to 150°C for 10 min, the same thermal history experienced by Blend 1 in Figure 3a. The micrograph is dominated by ~0.5 μm spherical depressions from which the material rich in HPB38 has been etched. That this indeed reflects liquid-liquid separation was confirmed observation of coarsening of the dispersed particles after aging in the melt. It should be noted that the volume fraction of the second phase, calculated by standard line segment analysis,[22] is $\phi_d = 0.08$. This is larger than the volume fraction $\phi_{HPB} = 0.05$ present in the blend, which implies some miscibility of HPB38 and PE. We have not calculated equilibrium conditions for this multicomponent system, but suspect that low molecular weight PE chains have entered the dispersed HPB-rich domains.

Slower crystallization of the thick CT samples yields interesting differences in morphology depicted in Figure 4. Unblended PE (not shown) has ~10 μm diameter unbanded spherulites characterized by lamellar bundles about 0.5 μm thick. Figure 4a shows slowly cooled Blend 1 to have a conspicuously nonuniform morphology. About half the material resembles PE with ~5 μm regions of fairly parallel lamellar bundles. The remaining portion is ~5 μm regions of thinner, poorly organized lamellar bundles about 0.2 μm thick. Blend 2 (not shown) appears the same, except that the "fine" regions have ~0.1 μm bundles. It should be recalled that these two blends contain only 10% HPB. The inference is that HPB with lower melting temperatures (Table I) has been rejected from slowly growing PE spherulites. Continuing this line of reasoning, the fine lamellar regions result from subsequent crystallization of liquid containing ~20% HPB. This second solidification process may

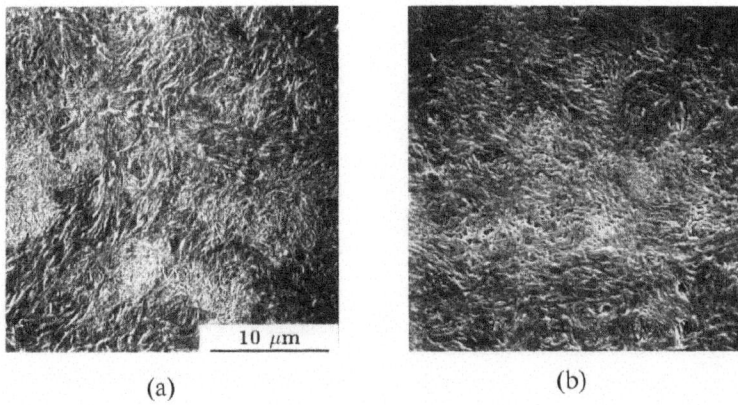

(a)   (b)

Figure 4. Micrographs (at equal magnification) of slowly cooled CT samples. (a) Blend 1; (b) Blend 3.

involve cocrystallization[23] or local segregation of HPB. The important point is that the presence of a small fraction of HPB chains has a profound effect on the morphology of the slowly cooled semicrystalline material. That should be contrasted to the quench crystallized morphology (Figure 3) which is insensitive to blending at the low concentrations employed here.

Figure 4b shows the morphology of the Blend 3 CT sample after slow crystallization from the two phase melt. One striking feature is that the volume fraction of the dispersed phase, $\phi_d \approx 0.04$, is about half that in Figure 3. It must be acknowledged, however, that the coarse spherulitic texture complicates observation of smaller domains. There has been aggregation of dispersed particles into "patches" which also contain irregular lamellar structures, though not so fine as those in slowly cooled Blends 1 and 2. It is reasonable that HPB-rich droplets with ~1 μm diameter would be moved together ahead of slowly advancing PE spherulite boundaries. The regions of irregular lamellae suggest rejection of less crystallizable chains during the first stages of crystallization, similar to the effect seen in Blends 1 and 2. No irreversible chemical changes were caused by molding and slow cooling, as the Blend 3 CT sample recovers the expected morphology after remelting and quenching.

A more extensive study, including calorimetry, of these model systems has been reported recently.[24] There it is concluded that there is partial miscibility of PE and HPB38 in Blend 3, and that rejection of low molecular weight PE chains during slow crystallization modifies the thermodynamics of the uncrystallized portion. Regardless of questions about quantitative aspects of Blend 3, the physical microstructures of CT samples for the three blends are understood. Blends 1 and 2 have structural nonuniformities which derive from slow solidification of a one phase melt containing two types of chains. Blend 3 contains a dispersed phase of very low crystallinity; even if this has some unbranched PE chains, the droplets are certainly much more compliant than the PE (or PE-rich) matrix.

Fracture Toughness

J integral data are presented in Figure 5 and in Table III. Fracture toughness J_C is obtained by extrapolating J to the value of *apparent* crack growth Δa, the steep

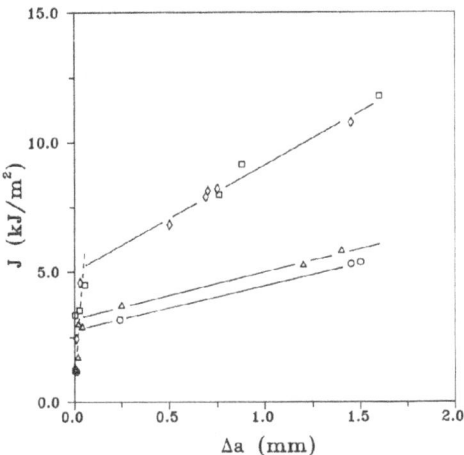

Figure 5. J integral versus crack growth Δa for PE (O), Blend 1 ($\Diamond$), Blend 2 ($\Box$) and Blend 3 (Δ).

Table III
Fracture Toughness Results

Blend	$\dfrac{\text{Avg. br.}}{1000\ \text{C}}$	$\rho\ [\text{kg/m}^3]$	$J_C\ [\text{kJ/m}^2]$
PE	0	967	2.7
1	1.5	957	5.2
2	6.1	954	5.2
3	5.9	959	3.3

dashed line in Figure 5, caused by stretching craze ligaments at the notch tip. Sample dimensions are adequate to conform to initiation and advance of the crack with J control under plane strain conditions:[21]

$$B > \frac{15J}{\sigma_y} \qquad (2a)$$

$$W\text{-}a_0 > \frac{15J}{\sigma_y} \qquad (2b)$$

In these experiments thickness B and ligament length $W\text{-}a_0$ are ≥ 10 mm, and the yield stress σ_y is greater than 20 MPa. Hence eqs. 3a,b are satisfied for $J < 13$ kJ/m^2.

Blends 1 and 2 are nearly twice as tough as PE, while Blend 3 has J_C only about 20% greater than PE. The conclusion we draw is that composition heterogeneity leads to enhanced fracture toughness, provided the mixture of chains was single phase in the melt. Rubbery second phase particles provide marginal toughening at best. Another interesting point is that the average amount of branching has no effect on J_C (Blends 1 and 2); the significant factor seems to be the relative number of differently crystallizable chains in the one phase melt which dominates the morphology established by slow crystallization. The unlike behaviors of Blends 2 and 3, each having the same average branch concentration, show again that this quantity has no fundamental relevance to toughness.

Some appreciation for the origin of enhanced fracture toughness in Blends 1 and 2 can be gained from inspection of the damage zones formed under plane strain conditions. At $J \approx 1$ kJ/m^2 a craze forms at the notch tip and grows in length l and height h with increasing deformation (increasing J) until fibrils at the root of the craze fail. This local fracture signals the onset of crack advance and defines J_C. When the crack is initiated (by craze breakdown) the craze has grown to $l > 500$ μm and $h > 20$ μm, so it is termed a *macrocraze*. Figure 6 presents macrocrazes in PE, Blend 1 and Blend 3. Oriented polymer within the craze appears as coarse fibrils in PE, and a continuous, porous web-like structure in Blend 1. Macrocrazes in Blend 2 (not shown) are web-like as in Blend 1, and those in the weaker, phase-separated Blend 3 are fibrillar like PE.

Thus plane strain fracture toughness in polyethylene and these model blends is essentially a measure of strength of the macrocraze formed in response to triaxial stresses at the notch tip. Web-like, as opposed to fibrillar, crazed material is correlated with increased fracture toughness. Similar observations were made by Huang and Brown[13] for macrocraze appearance and crack growth rates in commercial ethylene-hexene copolymers with increasing hexene content (though we suspect that the important parameter is composition distribution, not average branch concentration). Recall that the melt-miscible Blends 1 and 2 have nonuniform morphologies

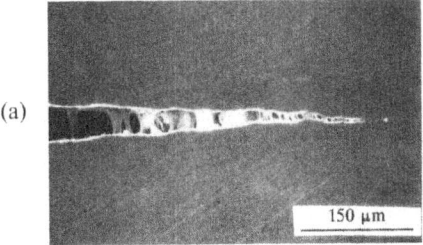

(a)

150 μm

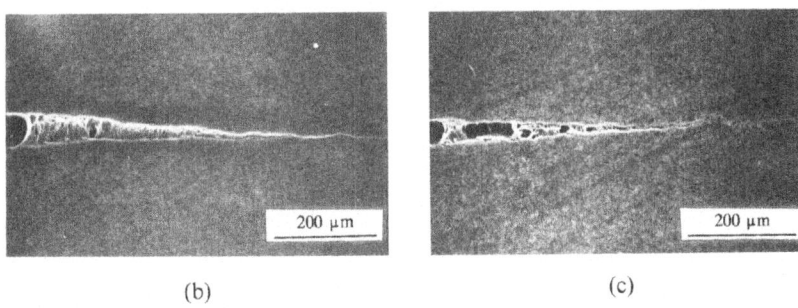

(b)   (c)

200 μm   200 μm

Figure 6. Transverse sections of CT samples after plane strain deformation. (a) PE, J = 3.2 kJ/m². Crack has grown to Δa = 0.25 mm; (b) Blend 1, J = 4.6 kJ/m²; Blend 3, J = 3.0 kJ/m².

when slowly cooled. Figure 7 illustrates one aspect of the relation between blend composition, structural heterogeneity and crazing. Toward the advancing tip of the damage zone one observes the early stages of craze development; in Figure 7 there are multiple crazes of height $h \approx 1$ μm. Not shown are micrographs of unblended PE, in which similar features appear to run with frequent branching from spherulite center to spherulite center. Craze initiation in Blend 2 is clearly favored in regions of fine lamellar bundles which are rich in HPB22 (Figure 7a). The phase separated Blend 3 has a comparable appearance in this early stage of craze growth (Figure 7b). One again sees crazes running preferentially through the "fine" regions, though the nonuniformity is not so pronounced as in Blend 2. Conspicuous by their absence are the dispersed particles - compare to the undeformed material in Figure 4. This

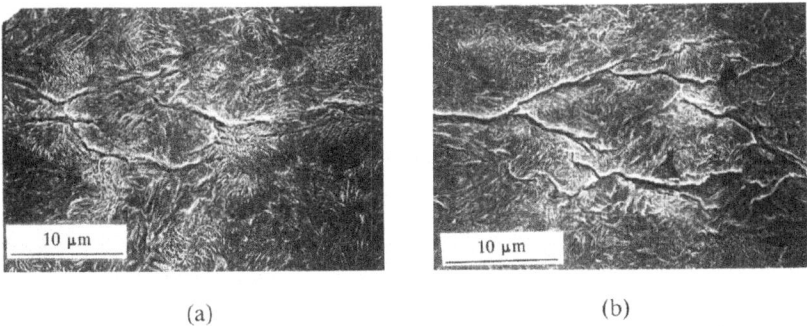

10 μm   10 μm

(a)   (b)

Figure 7. Etched surfaces of craze tip regions of CT samples. (a) Blend 2, J = 3.3 kJ/m²; (b) Blend 3, J = 3.0 kJ/m².

193

implies that crazes initiate at or near the rubbery droplets, which are no longer visible.

A plausible explanation is as follows. Craze initiation occurs at spherulite centers (PE) or in regions of high copolymer content (blends). The critical issue seems to be craze growth. In "one phase" blends there are substantial regions of fine lamellar bundles which transform into the nearly continuous web of oriented craze material (see Figure 8b). In PE or the phase separated Blend 3, the craze must grow by local drawing of conventional spherulites with coarse lamellar bundles. These evolve into fibrous macrocrazes (Figures 6a,c) which fail at smaller J. The underlying reasons for these differences are not known at this time. It is obvious, however, that a relatively small fraction of melt-miscible chains can alter substantially the morphology of slowly cooled blends and increase fracture toughness. The presence of a dispersed phase rich in amorphous chains has little effect on the morphology of the matrix or on fracture toughness of the blend.

Tensile Properties

Tensile testing (Table IV) was done on slowly cooled ($1°/min$) films which have higher densities (crystallinities) than the CT samples. By comparing densities in Tables III and IV, one sees that the cooling rate in CT samples was greater than $1°/min$. A general consequence of adding HPB to unbranched homopolymer is to lower the density. Tensile modulus E and yield stress σ_y of the blends then follow density/crystallinity in the same manner as unblended PE and copolymers.[25,26] The tensile properties of the blends are generally similar to one another. Note particularly that there is no difference between the "one phase" Blend 2 and the "two phase" Blend 3 with the same average branch concentration, though fracture toughnesses are quite distinct.

Table IV
Tensile Properties

Blend	ρ [kg/m^3]	E [MPa]	σ_y [MPa]
PE	975	1030	28.9
1	965	880	26.4
2	961	830	25.8
3	964	830	26.1

CONCLUSIONS

The major finding in this work is that a small amount (10%) of copolymer in a PE blend can substantially modify physical microstructure and plane strain fracture toughness. These changes are caused by liquid-solid phase separation during slow crystallization of melt-miscible chains having different amounts of branching. The similarity of Blends 1 and 2 shows that the number of different chains is more significant than branch concentration. It is well established that short chain branches lower the melting temperature by total or partial exclusion of branch points which decreases lamellar thickness and increases the (surface) free energy of the crystallites.[27,28] The present studies on well characterized model blends show that the amount of branching in the added component has only second order effects on morphology and properties, provided the melt remains single phase.

Branch concentration in the added copolymer can be increased to the point at which liquid-liquid phase separation occurs in the melt. The thermodynamic analysis employed here predicts correctly when two phases will (or will not) form, but not volume fractions and compositions of coexisting phases in the multicomponent system. We should point out that our approach to phase separation is a conventional mean-field one, and should not be confused with the quite different effects reported by Hill, et al.[29] The phase separated model Blend 3 has nearly the same toughness as unblended PE and the same tensile properties as Blend 2. Quantitative interpretation of Blend 3 behavior must await development of theory or experiment to establish compositions of the phase domains.

These results can be used to help understand commercial ethylene - α-olefin copolymers. Considering first "medium density polyethylene", we point first to its similarity to Blends 1 and 2; single phase in the melt, with the distribution skewed to the unbranched side. We postulate that the superior fracture resistance of commercial copolymers derives from the morphology established during slow cooling. One significant difference is that a commercial copolymer with, for example, an average of ~6 branches/1000 C atoms has a narrower distribution than model Blend 2. That will modify details of the nonuniform morphology shown here to be important in craze initiation and growth, which probably accounts for even greater toughness. Chung and Williams[30] have recently reported $J_C \approx 27$ kJ/m^2 at room temperature for a MDPE of unspecified branching. One prediction from this work is that the difference between PE and "one phase" model blends, and perhaps medium density copolymers, will vanish when the polymers are quench crystallized to suppress the formation of nonuniform morphology. Unfortunately it is very difficult to cool rapidly enough samples of dimensions required for plane strain testing.

Linear low density PE[1,2] (and more branched "very low density PE"[3]) is analogous in one sense to Blend 3 which is phase separated in the melt. An important feature of those "two phase" materials is that they have continuous (as opposed to ideally bimodal) composition distributions. Hence, when the distribution is wide enough to force separation of a minority phase of highly branched chains, the remaining "matrix" material is very heterogeneous and will crystallize something like Blends 1 and 2. In that case it becomes difficult to ascribe properties to the presence of a second phase, since the matrix is quite different from (unblended) PE. The present experiments show that a second phase of highly branched chains is insufficient to increase plane strain fracture toughness at room temperature. In commercial LLDPE, however, there is no doubt that significant toughening occurs at temperatures below -20°C.[2] The extent to which this is caused by the second phase or the attendant composition heterogeneity in the matrix phase is unresolved. We note in closing that a broader overall composition distribution will generally result in a larger volume fraction of phase separated material. Hence ϕ_d is an indirect measure of the distribution width, which may lead to apparent correlations between properties and the amount of second phase.

ACKNOWLEDGEMENTS

This work was supported by the Gas Research Institute, Physical Sciences Department (Contract 5090-260-2066). We are indebted to I.R. Harrison for the TREF characterizations. F. Mirabella is thanked for numerous discussions.

REFERENCES

1. F.M. Mirabella and E.A. Ford, *J. Poly. Sci. Polym. Phys. Ed.* **25**, 77 (1987).
2. F.M. Mirabella, S. Westphal, P.L. Fernando, E. Ford and J.G. Williams, *J. Polym. Sci. Polym. Phys.* **26**, 1995 (1988).
3. R.A.C. Deblieck and V.B.F. Mathot, *J. Mat. Sci. Lett.* **7**, 1276 (1988).
4. L. Wild, *Adv. Polym. Sci.* **98**, 1 (1990).
5. I.R. Harrison, Pennsylvania State University, private communication.
6. J.M Crissman, "Reference Standard Polyethylene Resins and Piping Materials", Gas Research Institute Final Report, GRI-86/00070 (1987).
7. B. Crist, H. Swei and S.H. Carr, in "Proc. 12th Plastic Fuel Gas Pipe Symposium", American Gas Assoc., Arlington, VA, Sept. 1991, pp. 480-485.
8. H. Swei, Ph.D. Thesis, Northwestern University (1991).
9. B.J. Bauer, *Poly. Eng. Sci.* **25**, 1081 (1985).
10. A. Nesarikar and B. Crist, unpublished results.
11. N. Brown, X. Lu, Y. Huang and R. Qian, *Makromol. Chem.* **31**, 711 (1991).
12. A. Lustiger and R.L. Markham, *Polymer* **24**, 1647 (1983).
13. Y.-L. Huang and N. Brown, *J. Polym. Sci. Polym. Phys. Ed.* **29**, 129 (1991).
14. S. Hosoda and A. Uemura, *Polymer J.* **24**, 939 (1992).
15. T. Krigas, J.M. Carella, M.J. Struglinski, B. Crist, W.W. Graessley and F.C. Schilling, *J. Polym. Sci. Polym. Phys. Ed.* **23**, 509 (1985).
16. J. Rhee and B. Crist, *Macromolecules* **24**, 5663 (1991).
17. P.R. Howard and B. Crist, *J. Polym. Sci. Polym. Phys. Ed.* **27**, 2269 (1989).
18. R.J. Roe and L. Lu, *J. Polym. Sci. Polym. Phys. Ed.* **23**, 917 (1985).
19. H. Swei, B. Crist and S.H. Carr, *Polymer* **32**, 1440 (1991).
20. L. Wild, Quantum Chemical/USI, private communication.
21. ASTM E813 in "Annual Book of ASTM Standards", American Society for Testing Materials, Vol. 03.01 (1987).
22. E.E. Underwood, "Quantitative Stereology", Addison-Wesley, Menlo Park, CA (1970), pp. 81 ff.
23. R.G. Alamo, R.H. Glaser and L Mandelkern, *J. Polym. Sci. Polym. Phys. Ed.* **26**, 2169 (1988).
24. J. Rhee and B. Crist, *J. Polym. Sci. Polym. Phys. Ed.*, submitted for publication.
25. B. Crist, C.J. Fisher and P.R. Howard, *Macromolecules* **22**, 1709 (1989).
26. A.J. Peacock and L. Mandelkern, *J. Polym. Sci. Polym. Phys. Ed.* **28**, 1917 (1990).
27. R. Alamo, R.Domszy and L. Mandelkern, *J. Phys. Chem.*, **88**, 6587 (1984).
28. S. Hosoda, *Polym. J.*, **5**, 383 (1988).
29. M.J. Hill, P.J. Barham and A. Keller, *Polymer*, **33**, 2530 (1992) and references therein.
30. N. Chung and J.G. Williams, in "Elastic-Plastic Fracture Test Methods: The User's Experience", ASTM STP 1114, Vol. 2, J.A. Joyce, ed., American Society for Testing Materials, Philadelphia, PA (1991), pp. 320-339.

TOUGHENING ENGINEERING THERMOPLASTICS WITH FUNCTIONALIZED ETHYLENE-PROPYLENE RUBBERS: IMPACT MODIFICATION OF STYRENE-MALEIC ANHYDRIDE COPOLYMER

S. Datta, N. Dharmarajan and G.Ver Strate

Polymers Group
Exxon Chemical Co.
Linden
New Jersey 07036

INTRODUCTION

This paper describes impact toughened two component blends of Styrene-maleic anhydride copolymer (SMA) with amine functionalised ethene-propene copolymer (EP-amine) and three component blends of SMA, EP-amine and high density polyethylene (HDPE). This is an alternate to the procedure of impact modifying SMA with copolymerised polybutadiene rubber or styrene-acrylonitrile copolymer grafted[1,2] polybutadiene rubber. These impact modified SMA compositions are deficient in their weatherability. Weatherability should be enhanced in our blends since we use a fully saturated polyolefin as the modifiers instead of a polyunsaturated polybutadiene rubber.

THEORY/CONCEPT

Blends of engineering thermoplastics (ETP) and polyolefins such as ethylene-propylene (EP) copolymer are normally incompatible, even in the melt. This arises principally from a difference in polarity between the highly polar engineering thermoplastics and the essentially nonpolar polyolefins. However blends of engineering thermoplastics and a minor amount (5% to 25%) of EP have been shown[3-5] to have significant improvement in mechanical properties, particularly in impact strength. This effect has been most thoroughly reported for polyamides[3], polyesters[4] and polycarbonates[5] but has been studied[6a,b] for Styrene-Maleic anhydride copolymer (SMA) for two cases with restrictive grafting procedures. The principal cause for this improvement in properties appears to be the stress distribution promoted by a fine dispersion of rubbery inclusions (particle size of 1 to 3 μm) in the ETP matrix.[7] Control of dispersion of polyolefins in blends with ETP is usually achieved through the use of compatibilizers which are block or graft polymers that act as interfacial agents. These polymeric compatibilizers are most conveniently formed by reaction of complementary functionality between the engineering thermoplastic and the polyolefin. The effectiveness of this toughening also depends on other factors such as rubber levels, and adhesion at the interface between the dispersed elastomer particles and the polymer matrix.[8] This paper shows that amine functionalised EP rubber (EP-amine) can be effectively used as an impact modifier for SMA. The reaction of amine groups in EP-amine with the maleic anhydride groups in the thermoplastic leads to the formation of the graft polymer which acts as a compatibilizer in this blend.

The use of a mixture of high density polyethylene (HDPE) and EP rubber as a impact modifier for polypropylene has been investigated by Stehling et. al.[9] The solubility parameters of the different components (calculated on a consistent basis) are polypropylene - 6.32, EP - 7.64 and HDPE 8.0 $cal^{0.5}.cc^{-0.5}$. Large differences in solubility parameter in uncompatibilized blends lead to correspondingly large interfacial tension; hence at thermodynamic equilibrium only two interfaces (a) polypropylene and EP and (b) HDPE and EP are favored. This ordering of the polymeric phases according to their solubility parameters leads to a morphology where crystalline HDPE particles are embedded in the EP particles: this combination of EP and HDPE forms the rubbery inclusions in the polypropylene matrix. We show that similar ordering of solubility parameters exist for mixtures of SMA, EP-amine and HDPE and leads to the preferred interfaces being between (a) SMA and EP-amine and (b) EP-amine and HDPE. The interfacial tension between SMA and EP-amine is lower than the solubility parameter calculations suggest since chemical reactions between these two polymers lower the interfacial tension. This favors the morphology indicated above. The morphology of these three component blends is similar to that described above and in effect the HDPE acts as a inert 'filler' in the EP-amine inclusions.

EXPERIMENTAL/RESULTS

SMA (available from Arco Chemical Co.) is a random copolymer of styrene and maleic anhydride containing typically less than 15% by weight of the second component. SMA resins are brittle and require impact modification for engineering applications. Two types of SMA were considered for our experiments: SMA-1 (Mw = 100 000, Maleic anhydride content 8%, available as Dylark 232 from Arco Chemical Co) and SMA-2 (Mw = 60 000, Maleic anhydride content 14%, available as Dylark 332 from Arco Chemical Co). EP-amine has a ethylene content 56 mol %, Mw = 120000, Mw/Mn = 2.1 and 7 mmoles of amine per 100 gms of polymer. The novel synthetic procedure which leads to direct formation of functionalised EP-amine has been described by us elsewhere.[10] The distribution of reactive, functional groups in the EP-amine polymer is, to the best of our understanding, random along the chain. The concentration of functional groups can be varied in our polymerization and functionalization procedure, however, we have concentrated on materials which have about 0.3 to 0.5 mol% of functionality. For a polymer chain of the above average molecular weight this corresponds to approximately 6 to 10 functional groups per chain. HDPE is available from Exxon Chemical Co.

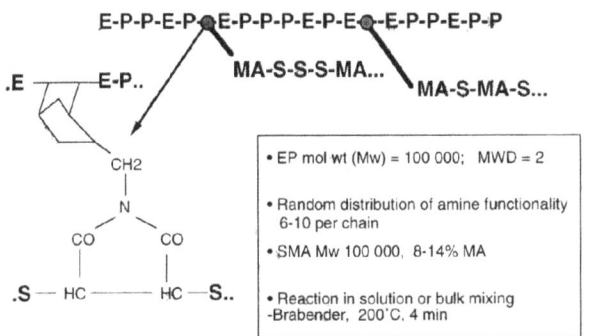

- • EP mol wt (Mw) = 100 000;   MWD = 2
- • Random distribution of amine functionality 6-10 per chain
- • SMA Mw 100 000, 8-14% MA
- • Reaction in solution or bulk mixing -Brabender, 200˚C, 4 min

Figure 1. Imidization reaction in the formation of SMA-g-EP-amine.

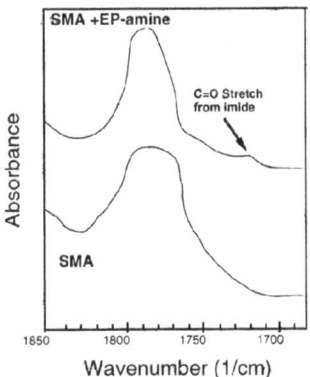

Figure 2. IR spectra for SMA and SMA-g-EP-amine

Formation of Graft Polymers Between SMA and EP-amine

The reaction of SMA and EP-amine is carried out either by prolonged reflux (6-12 hours) of a xylene solution of the two polymers or by the reaction of the neat polymers in the melt in a high intensity shear mixer (Brabender mixer) at about 180° C - 200° C for a few minutes. The reaction scheme is shown in Figure 1.

The progress of the reaction can be monitored by changes in the IR absorbances of the carbonyl region (1600 cm^{-1} to 2000 cm^{-1}) for the blend. For the compositions of SMA and EP-amine that we are using and at the blend ratios of approximately 80/20, respectively, of the above polymers, only 0.2 to 0.5% of the maleic anhydride residues are reacted. Thus the anticipated diminution of the maleic anhydride band at 1760 cm^{-1} is too small to be practically observed. However the reaction of EP-amine with the anhydride groups of SMA does lead to the formation of imide groups which|are|visible as separate and distinct absorbances in the IR spectra. An example of such a reaction product is shown in Figure 2 which shows the presence of a new carbonyl resonance at 1735 cm^{-1}, characteristic of imides, from the reaction of SMA and EP-amine.

Evidence that the chemical reaction outlined above does lead to the formation of EP-g-SMA polymers is obtained by a differential solubility experiment. In this procedure a thin pad of the polymer is successively extracted with refluxing acetone (which dissolves SMA) and hexane (which dissolves EP-amine). The soluble fraction from each extraction is isolated and the relative amounts of EP-amine and SMA, are analysed by IR. Absorption bands at 722 cm^{-1} and 1012 cm^{-1}, which are characteristic of EP and SMA, have been used for this determination. Blends of EP and SMA in the ratio of 50/50 by wt, where the EP is a nonfunctionalized polymer having 0.53 mol fraction of ethylene and a Mw (by GPC) of 139000 lead to a 'clean' separation of the polymers in this fractionation procedure. This is shown by the data in Figure 3 (EP-nonfunctional) where the fractionation of the blend of a nonreactive pair of polymers leads to a hexane soluble fraction which is all the EP in the blend and an acetone soluble fraction which is all of the SMA in the blend.

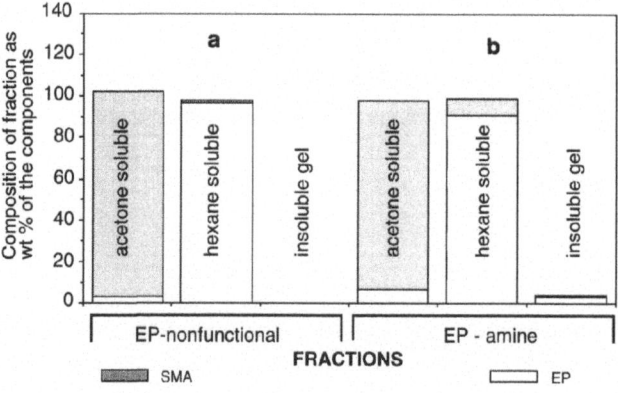

Figure 3. Extraction results of (a) a 50:50 blend of non functional EP and (b) a 50:50 blend of EP-amine and SMA.

Note the absence of (a) significant amounts of polymer having an intermediate solubility between these two extremes and (b) contamination of either the EP or the SMA fraction with SMA or EP, respectively. The situation is significantly different in the case of the reactive blend of EP-amine and SMA. The data in Figure 3 (EP-amine) shows the amount of polymer isolated in the hexane or acetone fraction is always a mixture of both SMA and EP-amine. A clear separation of the polymers is not achieved. The extent of cross contamination of the phases is about 5-7 wt%. This solubility property strongly suggests the formation of copolymers of SMA grafted onto an EP-amine backbone: the solubility of such polymers is intermediate between the two constituent polymers and is determined by the relative amounts of SMA and EP-amine. Some of the material does appear as insoluble gel: this is evidence for partial crosslinking.

a **Non-functionalized EP**   b   **Amine-EP**

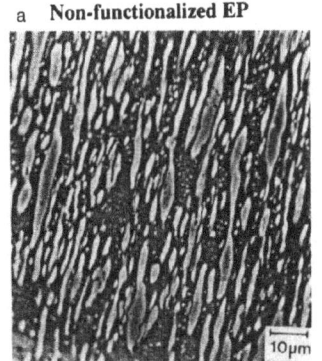 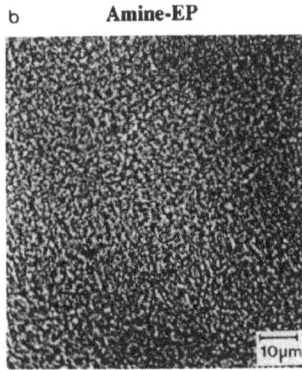

Figure 4. SEM micrographs of (a) a 70/30 blend of SMA and non functional EP, (b) a 70/30 blend of SMA and EP-amine

Two Component Blends - SMA And EP-Amine

Graft polymers formed by the reaction of EP-amine and SMA are active as compatibilizers. This effect is most effectively studied in blends of 90-70 % SMA and 10-30% EP-amine because of ease of microscopic determination of morphology.[11] The graft polymer is produced in situ during the blending of the SMA and the EP-amine in the Brabender mixer at 180⁰ C to 240⁰ C. Molded specimens of blends of nonfunctional EP and SMA in the above composition range show a very coarse morphology (Figure 4a) with the SMA as the continuous phase. The EP domains are elliptically distorted in the direction of the injection flow with the minor axis being about 5 to 8 μm while the major axis is about 20 to 40 μm. In contrast, blends of 90% SMA and 10% EP-amine (Figure 4b) show a very fine morphology where the EP-amine domains are approximately spherical with a diameter of about 1 μm. Blends in the composition of 70% SMA and 30% EP-amine show the same spherical EP-amine domains with a diameter of about 1 to 1.5 μm. Compositions containing significantly more EP-amine show cocontinuous EP-amine and SMA domains.

Table 1. Composition and Properties of SMA / Functional EP-amine Blends

Composition of Blends

SMA	100	90	85	80	75	70
EP-amine		10	15	20	25	30

Properties

Notched Izod (J/m)						
@ 21°C	16	44	60	131	230	295
@ -40 C	10	33	55	65	87	103
Flexural Modulus (Mpa)	3.08	2.81	2.32	2.22	2.01	1.71
Heat Distortion Temperature (°C) (@ 264 psi load)	94	92	-	88	-	87
Tensile Strength (MPa)	.055	.041	-	.031	-	.026

Blends of SMA and EP-amine show improvements in mechanical properties. Data is shown in Table 1 for room temperature notched Izod (ASTM D256, Method A), heat distortion temperature at 264 psi (ASTM D 1637), flexural modulus (ASTM D790, Method A) and tensile strength (ASTM D638) for these blends as a function of the composition (SMA/EP-amine). The data is compared to the corresponding data for neat SMA. Unmodified SMA is brittle and has a room temperature notched izod impact of about 16 J/m, which does not increase significantly in blends of nonfunctional EP and SMA. Notched izod impact strength of SMA modified with EP-amine at increases with increasing EP-amine levels. For example, at 25 wt % modifier level the impact strength is 230 J/m: this is a enhancement by a factor of 15. The impact enhancements of SMA modified with EP-amine is significantly higher when compared to other studies on SMA modification. Kim[2] who used SAN grafted to butadiene rubber as a modifier report impact improvements of only a factor of 4 at 25% rubber levels. The data for the change in impact strength at 25% EP-amine blended into either SMA-1 or SMA-2 with change in temperature (in the range 25°C to -40°C) is also shown in Figure 5. The impact strength of the blends slowly decreases with lower temperatures till the rubber modifier is essentially ineffective at -40°C. This is expected since that is the Tg for EP copolymers. Table 1 also shows the impact to stiffness (measured in terms of flexural modulus) balance of SMA modified using EP-amine. As expected with increasing EP-amine level there is a reduction in flexural modulus. With the addition of 25 wt % modifier to SMA, the flexural modulus drops from about 2.76 MPa to 1.72 MPa. Table 1 also illustrates the change in heat distortion temperature (HDT) of SMA resins after rubber modification. The high load heat distortion temperature (measured at 264 psi load) decreases from about 93°C for unmodified SMA to about 88°C at 25 wt% modifier.

Thermal properties of the polymer blends were measured using a DuPont 9000 differential scanning calorimeter. The samples were heated at 20°C / min from -100°C to 300°C. The glass transition temperature (Tg) of EP is -50°C while for SMA the Tg is 125°C. The blends show both the glass transition temperatures of the component polymers, which is typical of an immiscible polymer system.

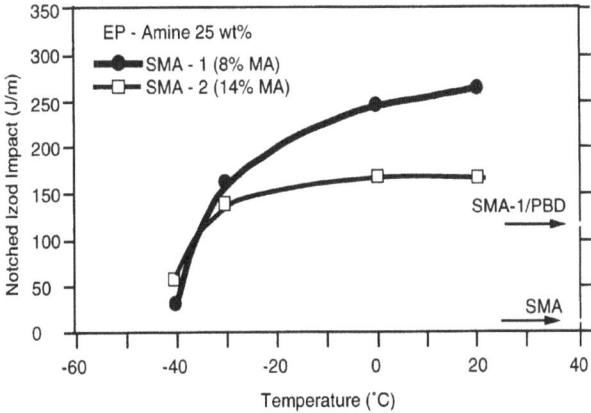

Figure 5. Impact strength of 25/75 blends of EP-amine with (a) SMA-1, (2) SMA-2 as a function of temperature. Note the impact strength at 23°C for SMA-1 and the commercially available polybutadiene modified SMA

Three Component Blends: SMA, EP-Amine, HDPE

Blends of these three components made by high intensity mixing at processing temperatures of 180°C to 220°C, which is above the melting point of HDPE have a single compatibilized polyolefin phase containing both EP and HDPE. Cooling these blends to ambient temperature preserves the phase boundaries but leads to preferential crystallization of the HDPE inside the polyolefin phase. HDPE should be completely enveloped by the functional EP polymer and not form an interface with the SMA matrix. The difference in solubility parameters between SMA and HDPE pair is about 1.1 cal$^{0.5}$cc$^{0.5}$ which is beyond the range of compatibility. By comparison, we anticipate that the interfacial energy of the EP-HDPE pair would be less than the SMA-HDPE pair.

These expectations about the morphology are verified by the SEM micrographs shown in Figure 6. The SMA-2/EP-amine(75/25) binary blend [Figure 6(a)] shows a fine dispersion of the EP-amine in the SMA-2 matrix with particle size in the range of 0.2 to 3.0 μm. Figure 6(b) shows the morphology of the ternary SMA-2/EP-amine /HDPE (75/15/10) blend. The light phase which appears as sub inclusion in the dispersed polyolefin phase phase is SMA-2. The polyolefin phase is a mixture of the amorphous EP-amine and crystallised HDPE. The lamellae of the HDPE are visible in the polyolefin phase as striations. The particle size of the dispersed phase which includes EP and HDPE ranges from 0.5 to 4 µm. Figure 6(c) illustrates the morphology of the SMA-2 - HDPE (75/25) binary blend system. As expected there is no compatibilization, and the HDPE particles are oriented in the direction of injection flow with domain sizes exceeding 30 µm along the major axis.

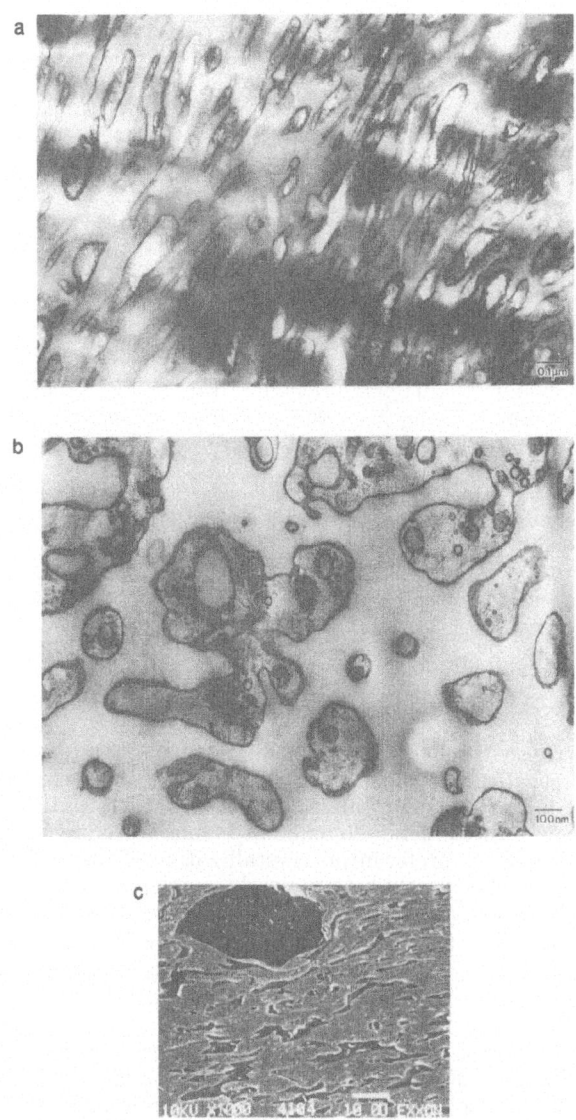

Figure 6. (a) TEM micrograph of SMA-2/EP-amine (75/25) blend; (b) TEM micrograph of SMA-2/EP-amine/HDPE (75/15/10) blend; (c) SEM micrograph of SMA-2/HDPE blend

The particle size distribution of the dispersed polyolefin phase in SMA was obtained from image analysis of the micrographs. Figure 7(a) shows the distribution of the functional EP (15 wt.%) in the SMA-2 matrix. The mean particle diameter is 1.13 μm and the majority of the particles are between 0.3 to 2 μm. This distribution is typical for most SMA/EP blend and does not appear to vary significantly with the type of SMA matrix. The particle size distribution for the SMA-2/EP/HDPE ternary system is shown in Figure 7(b). This appears to be similar to the SMA/EP blend with the slight exception that the ternary system has more large particles in the range of 3 to 6 μm. The mean particle diameter for the ternary system is 1.30 μm.

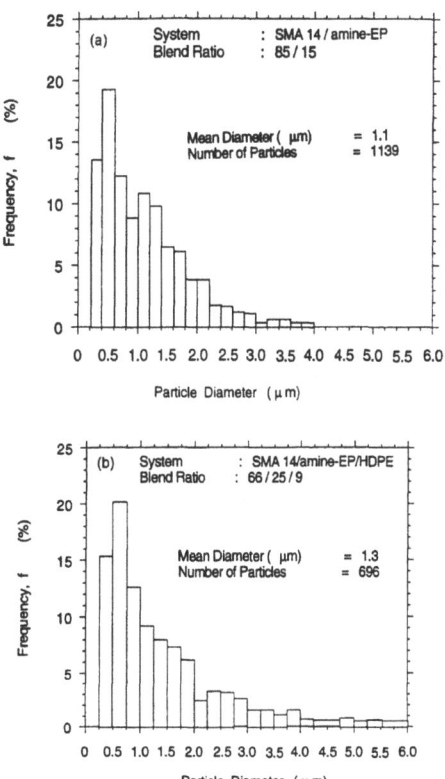

Figure 7. Particle size distribution for the dispersed polyolefin phase from SEM micrographs for (a) SMA-2/EP-amine (85/15) blends and (b) SMA-2/EP-amine/HDPE (66/25/9). Note the relative similarity in particle sizes irrespective of the amount or composition of the polyolefin modifier.

Table 2 summarizes the results from polyolefin area fraction analysis of the dispersed phase in the SMA-2 matrix. In the absence of HDPE sub-inclusions in the EP phase, this area fraction is the same as the volume fraction of the EP. In the three component SMA/EP-amine/HDPE blend the dispersed phase is now EP + HDPE and the inclusion of HDPE in EP causes a small volume increase of the rubbery inclusions. Consequently, the area fraction of the dispersed phase (0.34) is marginally higher than the EP volume fraction (0.29).

Table 2. Image Analysis results for the particle size distribution of the dispersed polyolefin phase of SMA-2 + EP-amine binary blend and SMA-2 + EP-amine + HDPE ternary blend

SMA-2	85	75	65
EP-amine	15	25	25
HDPE	0	0	9
Dispersed Phase	EP-amine	EP-amine	EP-amine + HDPE
Volume Fraction of polyolefin (composition)	0.18	0.29	0.34
Area fraction of polyolefin (image analysis)	0.18	0.30	0.34

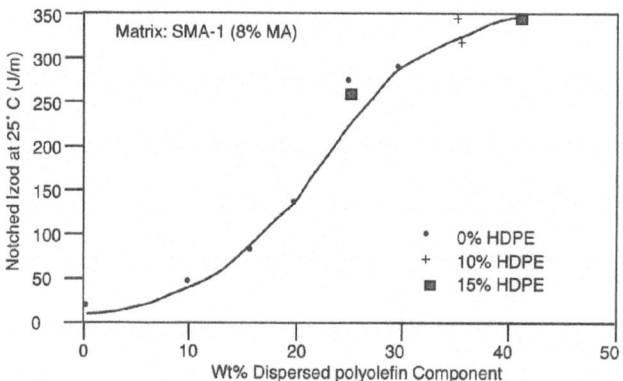

Figure 8. Notched Izod Impact strength of blends of SMA-1, EP-amine and HDPE

The impact properties of the ternary blends containing SMA-1 resin as the matrix are shown in Figure 8. The variation in notched izod impact strength is plotted as a function of the total weight fraction of the dispersed phase. The data for the ternary SMA/EP/HDPE and the binary SMA/EP system are coincident in the region (15 to 25% of polyolefin) of useful impact modification with polyolefins. The trend suggests that within the range of HDPE contents investigated in the three component mixture, it is only the total volume fraction of the dispersed phase that contributes to impact strength. This is a significant finding as it permits dilution of the functional EP with the relatively inexpensive HDPE component.

CONCLUSIONS

In conclusion we have demonstrated for the first time that EP-amine-g-SMA polymers which are effective compatibilizers for the SMA, EP-amine system by reaction of the amine (on EP-amine) and maleic anhydride (on SMA) functionalities. The synthetic approach is novel. These compatibilized blends are impact tough presumably due to presence of a stable dispersion of the rubber in the SMA matrix. The impact strength improvements that we have noted for these blends are better than those reported previously.[2] The particle size of the dispersed rubber phase in SMA for the best impact performance is about 1 to 1.5 μm. We have also demonstrated that it is possible to replace a part (<50%) of the EP-amine with HDPE without sacrificing impact toughening. These three component blends have a morphology where the particles of crystalline HDPE (~0.1μm in diameter) are embedded in the dispersed EP-amine phase (~1-3μm in diameter). Thus HDPE acts as a inert 'filler' in the EP-amine phase.

ACKNOWLEDGEMENTS

We thank Dr. E. N. Kresge, Dr. L. Kaufman and Mr. S. R. Wafalosky for helpful discussions and charitable criticisms. Our most sincere thanks are to them. We thank the management of the Polymers Group, Exxon Chemical Co., in particular Dr. J. J. O'Malley for his support of this work and for permission to contribute this paper. This work was done in its entirety at the Linden Technology Center, New Jersey.

REFERENCES

1. O.L.Stafford and J.J.Adams, U.S.Patent 3,642,949, 1972.
2. J.H.Kim, H.Keskkula and D.R.Paul, J. App. Poly. Sc., 40:183, 1990.
3. (a) U.S. Patent 4 174 358 (Du Pont), (b) U.S. Patent 4 536 541 (Du Pont), (c) U.S. Patent 4 594 386 (Copolymer), (d) U.S. Patent 4 339 555 (Mitsubishi), (e) U.S. Patent 4 346 194 (Allied Signal), (f) European Patent 291 796 (Bayer), (g) U.S. Patent 4 320 213 (Monsanto).
4. (a) U.S. Patent 4 172 859 (DuPont), (b) U.S. Patent 4 506 050 (Firestone), (c) German patent 3 306 008 (Huls).
5. (a) U.S. Patent 4 782 114 (Dexter), (b) U.S. Patent 4 550 138 (Uniroyal).
6. (a) S. Wu, Poly. Eng. Sci., 30:753 (1990), (b) U.S. Patent 4 742 116 (Stamicarbon)
7. (a) U.S.Patent 4 174 358 (DuPont), (b) U.S.Patent 4 172 859 (DuPont).
8. (a) C. B. Bucknall, Toughened Plastics, Applied Science, London (1977), (b) S. Wu, Polymer, 26, 1855 (1985).
9. F. C. Stehling, T. Huff, C. S. Speed and G. E. Wissler, J. Appl. Poly. Sci., 26: 2693, (1981).
10. S. Datta, in 'High Value Polymers' Ed A. H. Fawcett, Royal Society of Chemistry, London 1991
11. Morphology was determined for cryogenically (-190⁰ C) microtomed samples of the blend by light microscope followed by SEM, if necessary. Samples were stained with RuO_4 and contrast was achieved by Robinson back scattering.

STRUCTURE FORMATION DUE TO NON-EQUILIBRIUM LIQUID-LIQUID

PHASE SEPARATION IN POLYPROPYLENE SOLUTIONS

Hwan K. Lee,[a] Allan S. Myerson[a] and Kalle Levon[b*]

[a]Department of Chemical Engineering
[b]Department of Chemistry
Polytechnic University
6 Metrotech Center
Brooklyn, New York 11201

INTRODUCTION

The morphology of polymeric materials[1] is strongly influenced by liquid-liquid phase separation and crystallization. In general, there is a possibility that the liquid-liquid phase separation process is coupled with the crystallization process. The equilibrium state of semicrystalline polymer-solvent mixtures is presented by depiction with the phase diagrams as shown in Figure 1. The liquid-liquid phase separation region is located above and below the equilibrium liquid-solid transition with poor and good solvents, respectively. The cooperation of two phenomena which both depend on the non-equilibrium conditions may produce new structures, which are important from both scientific and industrial points of view.

Although these coupled phenomena are expected to occur in crystallizable polymer solutions, the influences of liquid-liquid phase separation on the formed morphologies were not studied extensively. Inaba et al.[2] found that the crystallization process could lock-in the modulated structure obtained by liquid-liquid phase separation in polymer blends. Schaaf et al.[3] observed globule-like morphologies resulting from liquid-liquid phase separation in polyethylene solutions. In both systems, however, the binodals are located above the melting points.

When the binodal exists below the melting point, the liquid-liquid phase separation has no equilibrium significance, but it may significantly affect the final morphology in processing. Liquid-liquid phase separation would be able to precede crystallization upon rapid quenching due to the high nucleation barrier to polymer crystallization.

In a previous paper[4] we reported the effects of interaction and polymer molecular weight on the phase transitions using isotactic polypropylene (i-PP) solutions with a series of phthalates. Dialkylphthalates with a different number of carbon atoms in the

---

*To whom the correspondence should be addressed.

New Advances in Polyolefins, Edited by
T.C. Chung, Plenum Press, New York, 1993

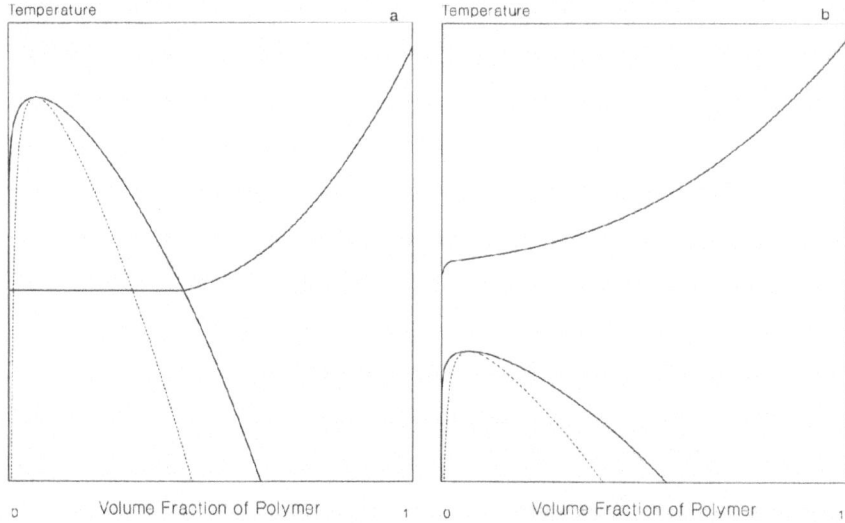

Figure 1. Two possible types of phase diagrams for semicrystalline polymer-solvent mixtures: (a) in a poor solvent; (b) in a good solvent. The binodal and the melting point curves are drawn by solid lines. The spinodal curves are drawn as dashed lines.

alkyl chain, were used to change the interaction between polymer and solvent. It was found that the liquid-liquid phase transition temperature decreased systematically as the solvent power increases and that the liquid demixing could occur far below the melting point with a favorable interaction (in the phthalate with a long alkyl chain). It was shown that an atactic polypropylene (a-PP) could be used to probe a non-equilibrium liquid-liquid phase separation, which exists in a low temperature region in an i-PP solution.

In this article we investigate the structure formation by applying a temperature-jump experiment to the i-PP solutions in which liquid-liquid phase separation and crystallization may occur simultaneously. Thermal quench conditions will be varied in order to control systematically the extents of liquid-liquid phase separation and crystallization. Optical Microscopy (OM) and Scanning Electron Microscopy (SEM) are used to observe the morphologies formed.

EXPERIMENTAL

Materials

Isotactic polypropylene (i-PP) with M_w 3.1x10^5 and M_w/M_n 3.4 was obtained from Himont R&D center. The solvents were series of 1,2-dialkylphthalates with the different number of carbon atoms in the alkyl chain, designated as C6, C7, C8, C8b, C9 and C10. Molecular characteristics such as the type of alkyl group and molecular mass, boiling point and supplier are summarized in Table I. All the solvents were used as received without purification.

Table 1. Molecular characteristics, boiling point and supplier of phthalates.

symbol	alkyl group	Mw	boiling point[a]	supplier
C6	hexyl	334	$210°C/5mmHg$	Exxon
C7	heptyl	362	$220°C/5mmHg$	Exxon
C8	octyl	391	$230°C/5mmHg$	Exxon
C8b	2-ethyl-hexyl	391	$390°C/760mmHg$	Aldrich
C9	nonyl	419	$252°C/5mmHg$	Exxon
C10	decyl	447	$256°C/5mmHg$	Exxon

[a]from the information provided by the suppliers.

Optical Microscopy with Temperature-Jump Experiment

i-PP films with a thickness of 20~40 μm were prepared according to the following procedure: i-PP powder was dissolved in hot decalin containing 2,6-di-tert-butyl-4-methylphenol (0.5% w/w on polymer) under nitrogen to form 0.7~1.0% w/w solutions; temperatures were close to the solvent boiling point; the solutions were held under reflux for 30 *min* after dissolution; the hot solutions were quenched by pouring them into an aluminum tray in an ice-water bath; the bulk of the solvent was allowed to evaporate in a current of air under ambient conditions and a transparent film was obtained; the residual solvent was removed by extraction with methanol; the films were dried in vacuum. Solutions of known concentration were prepared in flatcavity microslides. A cover glass was placed over the specimen and sealed in position with Conap Easypoxy Resin (Conap Inc., Olean, NY) in a nitrogen atmosphere. Care was taken to avoid contact of the specimen with the resin.

Structure evolution due to phase separations of the solutions was investigated by optical microscopy. The prepared solutions were heated slowly to 170°C and heating was continued for 10 *min* to ensure the complete dissolution. The samples were quickly transferred to a preset temperature in a Mettler hotstage. The time for equilibrating the specimen at the experimental temperatures varied from 30 to 240 *s* depending on the quench depths. The temperature for observation was determined to control systematically liquid-liquid phase separation and crystallization as is described in the discussion part. Condensation of the solvents during quenching was negligible for C8, C8b, C9 and C10. In case of C6, the bulk concentration of the solution was destroyed by the quenching process and the discussion of structure formation in C6 phthalate is omitted.

Scanning Electron Microscopy

The specimen for SEM were prepared by using DSC (Perkin Elmer DSC-7). i-PP powder and solvent were weighed to obtain 50~60 $\mu\ell$ in a stainless steel pan and dissolved at 170°C for 10 *min*. The specimen was cooled at a rate of 200°C/*min* to a set temperature and held 5 *min* to give rise to phase separation. Evaporation of the solvent was neglible during DSC running.

The sample was immersed in hexane for 48 hrs to extract the solvent at a room temperature. The solvent-free samples were dried in a vacuum oven for more than 2 days. The final sample was fractured in liquid nitrogen. The fractured surface was vacuum metallized with platinum using a sputter coater (Model 3, PELCO) and observed through a scanning electron microscope (Model 1200, AMR). Magnification ranged from 100x to 3,000x depending on the purpose of viewing.

RESULTS AND DISCUSSION

Experimental Scheme To Control Phase Separations

While the mechanism of crystallization is nucleation and growth (NG), for liquid-liquid phase separation the mechanism depends on the thermodynamic stability of the system. In the unstable region the phase separation occurs according to spinodal decomposition (SD) and in the metastable region according to NG.[5] Since the binodal curves merge with the spinodals at the critical point in phase diagram, one can avoid the NG mechanism for liquid-liquid phase separation during the cooling process through the critical concentration. Hence we prepared the i-PP solutions with the critical concentrations, which were around 2.5 wt%[4] with the various phthalates (C6, C7, C8, C8b, C9 and C10) to obtain SD as a liquid-liquid phase separation.

It was shown in the previous study[4] that atactic polypropylene (a-PP) can be used to probe non-equilibrium liquid-liquid phase separation of i-PP solutions which exists in a low temperature region. Since the a-PP used was purely amorphous, the liquid-liquid phase separation of a-PP solutions did not interfere with crystallization. The upper critical solution temperatures (UCST) of the i-PP solutions with less favorable interaction (in C6, C7 and C8) are shown in Figure 2. The UCST of the system with favorable interaction (in C8b, C9 and C10) may be estimated from the observation with a-PP solutions. The difference of UCST between a-PP and i-PP systems is approximately 3.0°C in C6, C7 and C8 phthalates in Figure 2. Applying an analogy to better solvents, one obtains the following values of UCST; 110°C for C8b, 92°C for C9 and 79°C for C10.

In order to understand the competitive phase separations, it is necessary to control

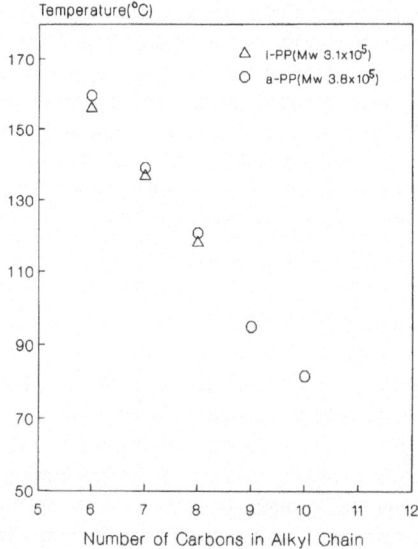

Figure 2. UCST of the i-PP and the a-PP solutions as a function of solvent quality (number of carbons in the alkyl chain of phthalate), determined by optical microscopy at a cooling rate of 2°C/min.

liquid-liquid phase separation and crystallization systematically. An approach to study this problem is to control the thermodynamic conditions through selection of the solvent while keeping the thermal conditions constant. The magnitude of the driving force for liquid-liquid phase separation can be varied without changing significantly that for crystallization because the liquid-crystal equilibrium is not sensitive to solvent quality. In addition, application of the different thermal quench condition to a given system will allow a more detailed analysis of the crystallization process coupled with liquid-liquid phase separation. Table 2 represents the thermal conditions applied to the i-PP solutions to change the extents of liquid-liquid phase separation and crystallization systematically. The melting points at the critical concentration were around 151~149°C

Table 2. Thermal conditions applied to i-PP solutions of the critical concentration. The number in brackets represents $\Delta T_{L-L}(^{\circ}C)$, the difference between UCST and the experimental temperature.

UCST	156	137	118	110	92	79
Temp.	C6	C7	C8	C8b	C9	C10
151	$\Delta(5)$					
140	$\Delta(16)$	^{a}X				
132		$^b\Delta(5)$				
121		$\Delta(16)$	X			
113			$\Delta(5)$	X	X	
100			$\Delta(18)$	$\Delta(10)$	X	X
85			$\Delta(33)$	$\Delta(25)$	$\Delta(7)$	X
70	$\Delta(86)$	$\Delta(67)$	$\Delta(48)$	$\Delta(40)$	$\Delta(22)$	$\Delta(9)$
50					$\Delta(42)$	$\Delta(29)$
0					$\Delta(92)$	$\Delta(79)$

[a]X crystallization without liquid-liquid phase separation.
[b]Δ crystallization coupled with liquid-liquid phase separation.

depending on the solvent power. One can control UCST with the different phthalate as shown in Table 2. In a given solvent, the mark X represents the experimental temperature above the UCST (crystallization without liquid-liquid phase separation) and the mark Δ does below the UCST (crystallization coupled with liquid-liquid phase separation). The number in brackets in Table 2 stands for ΔT_{L-L} ($^{\circ}C$), the difference between UCST and the experimental temperature. The other thermal quench conditions will be mentioned in the text. Thus we change ΔT_{L-L} without changing ΔT_{cry}, the difference between the melting and the experimental temperature too much at a certain temperature by selecting the proper phthalate. Both ΔT_{L-L} and ΔT_{cry} are increased by decreasing the experimental temperature in a given solvent. Combination of solvent selection and choosing the proper temperature allows the variation of ΔT_{cry} keeping ΔT_{L-L} constant.

Crystallization Without Liquid-Liquid Phase Separation

This part is devoted to the investigation of the spherulitic crystallization without liquid-liquid phase separation. It may give understanding to identify an origin of the various morphologies observed in the simultaneous phase separation. The possibility of liquid demixing on cooling is excluded based on the experimental phase diagram (corresponding to the marker **X** in Table 2).

The volume-filling spherulite formed at $100^{\circ}C$ for the i-PP/C10 solution are shown in Figure 3. The spherulitic formation was complete instantly by impingement on quenching. Although a slight increase in crystallinity was observed by the increased light intensity in crossed polars with duration of time, there was no appreciable change in structure. The crystallization of the i-PP/C7 pair at $140^{\circ}C$, however, did not produce the volume-filling spherulites. The crystal growth rate is very slow at $140^{\circ}C$ and the solvent molecules would be rejected rapidly to the growth front of crystal. As crystallization occurs at the site of nuclei, the nucleus-free region is depleted of the polymer and eventually remains out of polymer. When crystallization occurs at $100^{\circ}C$, the solvent may be entrapped in the interlamellar region due to rapid growth of crystal and slow diffusion of solvent. Then crystallization produces the continuous structure by impingement of growing spherulites.

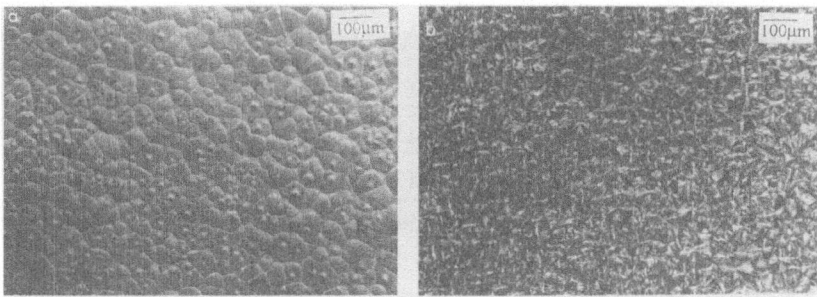

Figure 3.(a) Modulation-contrast micrograph and (b) polarized light micrograph for the 2.5 *wt%* i-PP/C10 solution quenched to $100^{\circ}C$.

The crystallization in a certain temperature range shows a dendritic growth. Figure 4 presents the spherulitic structure including dendrites at a late stage of crystallization. Spherulites were formed in the i-PP/C8 solution crystallized at $125^{\circ}C$ and dendrites started to appear at the front of a spherulite in 90 *min*. In the i-PP/C7 system crystallized at $132^{\circ}C$, a dendritic formation starts to grow from the edge of spherulites in 300 *min*. In case of i-PP/C8 crystallized at $115^{\circ}C$, the structure formation is dominated by a dendritic growth as shown in Figure 5. Thus a dendritic growth takes place at the earlier stage of crystallization with the higher supercooling; in 300 *min*. at 132°C, in 90 *min*. at 125°C and instantly at 115°C. The dendrite formation can be

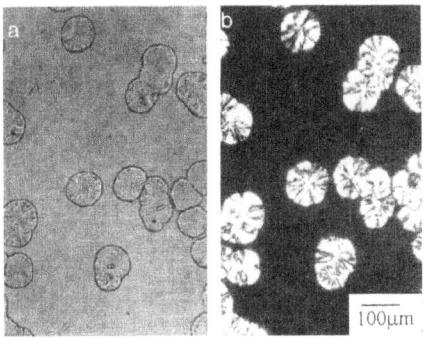

Figure 4. (a) Modulation-contrast micrograph and (b) polarized light micrograph for the 2.5 *wt%* i-PP/C7 solution crystallized at 132°C.

Figure 5. (a) Modulation-contrast micrograph and (b) polarized light micrograph for the 2.5 *wt%* i-PP/C8 solution crystallized at 115°C.

related to the polymer concentration gradient at the growth front of crystal. As Keith and Padden[6] suggested, noncrystallographic branching (dendritic formation) should be observed when the value of δ (=D/G, where D is the diffusion coefficient of the solvent molecule and G is the radial growth rate of the spherulite) is commensurate with the size of disordered region on the crystal surface. This may correspond to the situation of i-PP/C8 at 115°C.

The i-PP/C8b system at 113°C exhibits an interesting feature of crystallization shown in Figure 6. The crystallization occurs with the spherulitic formation in an early stage and is followed by a dendritic formation. In the region of high density of nuclei, the spherulites impinged each other to lock-in the structure without dendritic formation. In the low density region of nuclei, however, the crystal grows continuously to form a branched structure as seen in Figure 6. The compactness of the spherulite (or dendrite) seems to decrease with the growth of crystal and the growth front of the dendrite in a late stage shows the reminiscence of the impingement of spherulite. Thus a pseudo-volume-filling spherulite is produced.

Figure 6. (a) Modulation-contrast micrograph and (b) polarized light micrograph for the 2.5 wt% i-PP/C8b solution in the region of a low density region of nuclei, crystallized at 113°C.

According to the thermal conditions applied in Table 2 based on the experimental phase diagram, we should be able to observe liquid-liquid phase separation in i-PP/C7 at 132°C (ΔT_{L-L}=5°C) and in i-PP/C8 at 115°C (ΔT_{L-L}=3°C). The evidence of liquid demixing in the two systems was not clearly observed in Figure 4 and Figure 5, respectively. The undercoolings applied for the liquid-liquid phase separation may be too small to influence the crystallization process. The induction period for the thermal equilibrium is partly responsible; the polymer concentration of the liquid phase can be lowered by nuclei formed prior to reaching the UCST during cooling, resulting in a lower undercooling.

Crystallization Coupled With Liquid-Liquid Phase Separation

When crystallization is coupled with liquid demixing, the various structures are created by the competition between the two non-equilibrium processes and the final structures are strongly dependent on their dynamics. We investigate first the effects of a small extent of liquid-liquid phase separation on the spherulitic formation by optical microscopy in the i-PP/C8 system quenched to $113^{\circ}C$. The thermal condition corresponds to a deep quenching for crystallization ($\Delta T_{cry}=T_m-T=39^{\circ}C$) and a shallow quenching for liquid-liquid phase separation ($\Delta T_{L-L}=UCST-T=5^{\circ}C$). It is expected that the phase transitions should be dominated by crystallization. The droplets formed due to liquid-liquid phase separation are elongated to the radial direction of the spherulite as shown in Figure 7. The droplets are observed only inside spherulites suggesting that the droplet formation plays a role in nucleation for crystal growth. It is thought that the droplets are formed by concentration fluctuation in the thermodynamically unstable liquid state and subsequently split into the radial direction by a growing crystal.

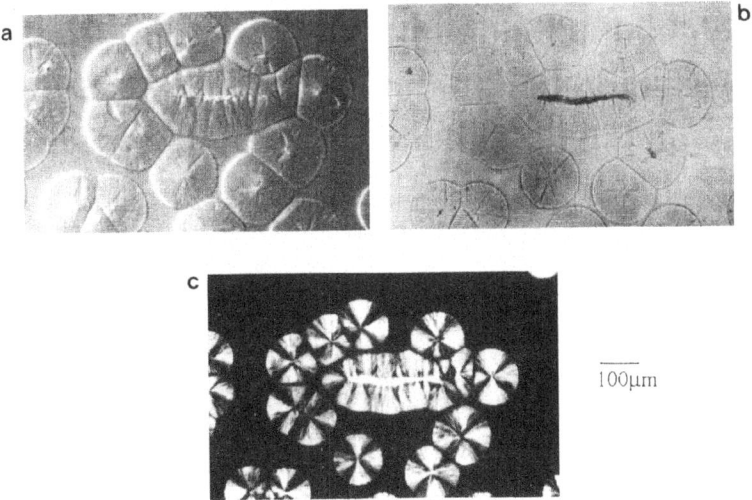

Figure 7. (a) Modulation-contrast micrograph, (b) transmitted light micrograph and (c) polarized light micrograph for the 2.5 *wt%* i-PP/C8 solution crystallized at $113^{\circ}C$.

When the i-PP/C8 solution is quenched to $100^{\circ}C$, the ribbon-like structure (the elongated droplets) is ubiquitous and the modulated structure produced by spinodal decomposition (SD) appears as shown in Figure 8. According to the linearized theory of SD[7,8], the wavelength of concentration fluctuation decreases with the increased undercooling for SD and the period at which the linearized theory is valid becomes short (rapid coarsening). Thus the modulated structure is obtained with the increased undercooling and subsequently frozen-in by the spherulitic crystallization prior to coarsening. The region at which the ribbon-like structure exists possesses a higher light intensity under crossed polars as seen in Figures 8 b and c, which were taken on the same field on the same sample, suggesting that the crystallization is enhanced by the

Figure 8. (a) Transmitted light micrograph, (b) modulation-contrast micrographs and (c) polarized light micrograph for the 2.5 $wt\%$ i-PP/C8 solution crystallized at 100°C.

droplet formation. Thus the solvent molecules rejected preferentially by growing crystals are piled up in the droplet region. This is supported by scanning electron microscopy as will be discussed. The ribbon-like structure, observed after extracting the solvent, occupies the vacant site oriented in the radial direction of a spherulite.

A different pattern formation is observed in the i-PP/C7 quenched to 121°C. The ΔT_{L-L} is raised to 16°C and the ΔT_{cry} is reduced to 31°C under the thermal condition applied. Therefore liquid-liquid phase separation in i-PP/C7 at 121°C is more competing with crystallization than in i-PP/C8 at 113°C. Although the overall

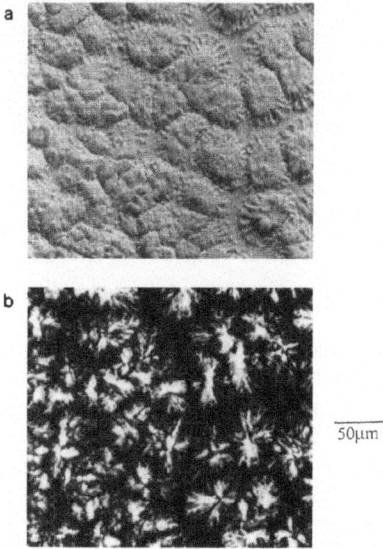

Figure 9. (a) Modulation-contrast micrograph and (b) polarized light micrographs for the 2.5 $wt\%$ i-PP/C7 solution crystallized at 121°C.

218

morphology is governed by the distribution of nuclei for crystallization, a spherulite contained the multiple droplets and again the droplets were oriented to the crystal growth direction. The droplets in a certain region are segregated to the growth front of a spherulite, as shown in Figure 9, indicating high mobility of liquid droplets and slow growth of crystal at the elevated temperature.[9]

The morphology obtained from non-equilibrium liquid-liquid phase separation under the deep quench condition was explored in the following step. The optical micrographs for the i-PP/C10 and i-PP/C9 pairs quenched to $0°C$ are shown in Figure 10. An uniform network structure was produced on quenching and frozen-in immediately. While the size of the spherulite is not clearly recognized from the polarized optical micrographs, we were able to determine the size of the domains due to liquid-liquid phase separation and spherulite by the depolarized light scattering pattern.[10] The structure formed consisted of small spherulites ($\sim10\mu$m) and large liquid phase separated domain ($\sim30\mu$m) in the i-PP/C10 solution crystallized at $0°C$. The size of liquid droplets is greater in i-PP/C9 ($\Delta T_{L-L}=92°C$) than in i-PP/C10 ($\Delta T_{L-L}=79°C$) as seen in Figure 10 a and b. Thus the droplet size is increased with the large extent of liquid-liquid phase separation under the same supercooling for crystallization ($\Delta T_{cry}=150°C$). The observed morphology is resulted from the rapid liquid demixing and locking-in the structure by crystallization. Although both types of non-equilibrium processes are extremely rapid, the coarsening rate of liquid-liquid phase separation is high enough to produce a larger domain due to liquid-liquid phase separation than the size of a spherulite which is determined by a competition between the nucleation rate and subsequent crystal growth. This is manifested from the fact that the droplet size in i-PP/C9 is greater than in i-PP/C10.

A similar trend is followed for the i-PP solutions in the series of phthalates, when crystallized at the different temperatures. Figure 11 shows the structure formed at $70°C$ for i-PP/C10 ($\Delta T_{L-L}=9°C$), i-PP/C9 ($\Delta T_{L-L}=22°C$), i-PP/C8 ($\Delta T_{L-L}=48°C$) and i-PP/C7

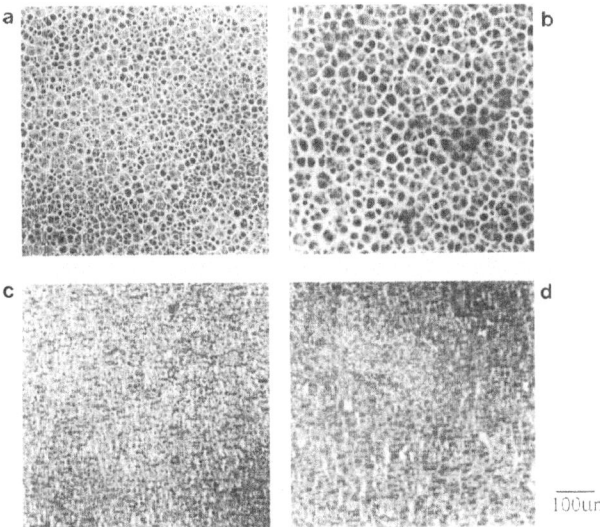

Figure 10. (a) and (b) Transmitted light micrographs and (c) and (d) polarized light micrographs for the 2.5 wt% i-PP/C10 solution (a and c) and for the 2.5 wt% i-PP/C9 solution (b and d), quenched to $0°C$.

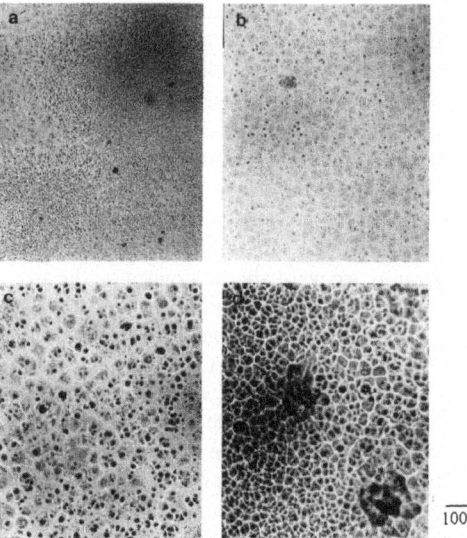

Figure 11. Transmitted light micrographs for the 2.5 *wt%* i-PP/C10 solution (a), the 2.5 *wt%* i-PP/C9 solution (b), the 2.5 *wt%* i-PP/C8 solution (c) and the 2.5 *wt%* i-PP/C7 solution (d), quenched to 70 °C.

($\Delta T_{L-L}=67^{\circ}C$). The size resulting from liquid-liquid phase separation increases with the increased undercooling for liquid demixing under the same supercooling for crystallization (note the liquid-crystal equilibrium is not sensitive to solvent power). The results indicate that the extent of liquid-liquid separation is controlled through the proper selection of solvent and the resulting structure is locked-in by crystallization.

The effects of temperature on the structure formation in the same solvent system can be found in Figure 10 (0°C) and Figure 11 a and b (70°C) for the i-PP/C10 and i-PP/C9 pairs. The size of a domain increases at a lower temperature (with the increased ΔT_{L-L}) in the both systems.

Scanning Electron Microscopy

The specimens were prepared by DSC. The samples were cooled from 170°C to a set temperature (0°C and 70°C) at a rate of 200°C/min and held 5 *min* to give rise to phase separation. Since the specimen of 2.5 *wt%* i-PP solutions tended to collapse during the solvent extraction process, 10.0 *wt%* i-PP solutions were also prepared for comparison.

Figure 12 shows the i-PP/C8b (2.5 *wt%*) system cooled to 0°C. The fine lacy structure (~3μm) and spherulites (~100μm) are observed indicating that spinodal decomposition occurred on cooling. It is clearly seen with high magnification that the interwoven structure and the void are bicontinuous and comprised of a small bead and a few ten micron entity resulting from the coarsening process. The spherulite size is much larger than observed in optical microscopy, suggesting that the spherulites are formed before the temperature reach to 0°C.

The morphology of the C8 system cooled to 70°C of 10 *wt%* polymer concentration shows the perforated spherulites (Figure 13). The peculiar structure is attributed to the liquid-liquid phase separation coupled with crystallization. The liquid droplets are formed by liquid-liquid phase separation on cooling and subsequently

Figure 12. Scanning electron micrographs for the i-PP/C8b system (2.5 *wt%*) cooled to 0°C at 200°C/*min* and held for 5 *min*.

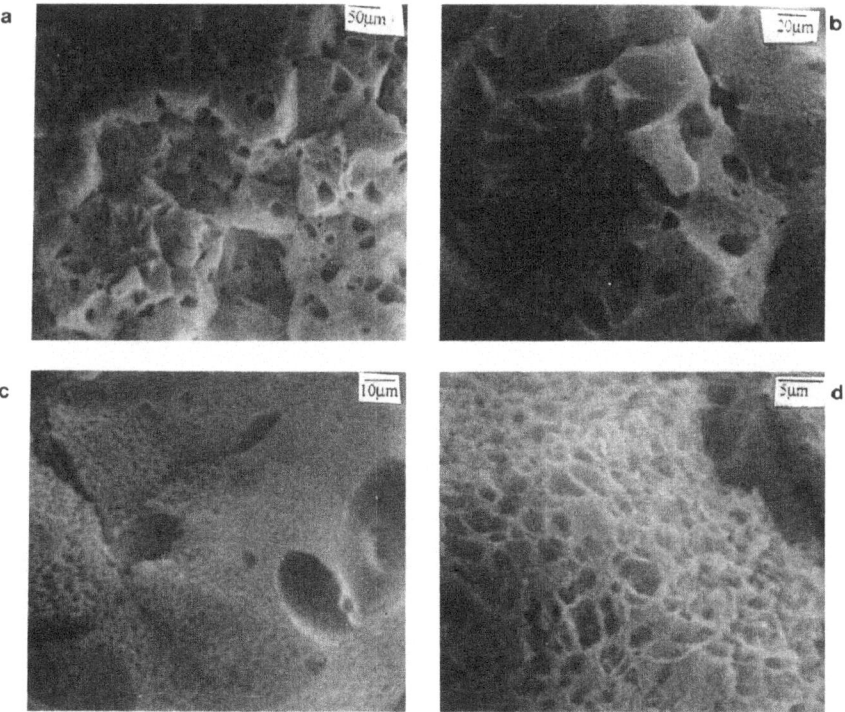

Figure 13. Scanning electron micrographs for the i-PP/C8 system (10.0 *wt%*) cooled to 70°C at 200°C/*min* and held for 5 *min*.

elongated in the radial direction of a spherulite by growing crystals. The droplets would grow by joining the solvent rejected by crystallization. This feature can be interpreted in conjunction with the ribbon-like structure formation observed with optical microscopy. The cellular structure observed with a high magnification (Figure 13 d) indicates that the liquid-liquid phase separation occurred during cooling according to nucleation and growth regime since the concentration of the specimen is in the off-critical region.

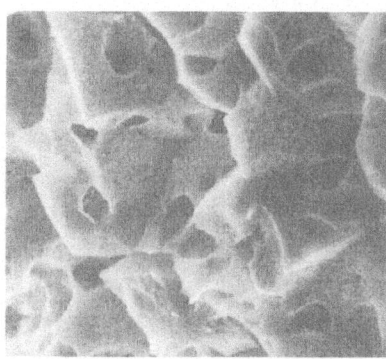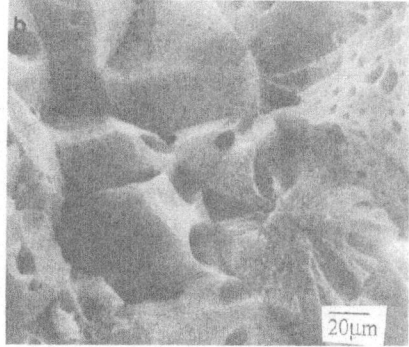

Figure 14. Scanning electron micrographs for the 10.0 *wt%* i-PP/C8b (a) and for the 10.0 *wt%* i-PP/C8 (b), cooled to 0°C at 200°C/*min* and held 5 *min*.

The scanning electron micrographs for the i-PP/C8b (10.0 *wt%*) and the i-PP/C8 (10.0 *wt%*) pairs cooled to 70°C are shown in Figure 14. The similar perforated spherulites are observed. The effects of temperature seems to be negligible due to a slow cooling in DSC.

CONCLUSIONS

The structure formation by liquid-liquid phase separation and crystallization in isotactic polypropylene (i-PP) solutions was achieved by a temperature-jump experiment. A series of dialkyl phthalates, with a different number of carbon atoms in the alkyl chain, was used to control the interaction between polymer and solvent. Various thermal quench conditions were applied to the i-PP solutions to control systematically liquid-liquid phase separation and crystallization. Firstly, when crystallization occurs in a temperature range above the upper critical solution temperature (UCST), the spherulitic growth is followed by the dendrite formation. The dendrites start to appear at the earlier stage of crystallization with the higher supercooling. Secondly, during the simultaneous phase transitions, the slow crystallization elongates the liquid droplets in the radial direction of a spherulite. Finally, a rapid crystallization locks-in the growth of liquid-liquid phase separation, in systems where the UCST is above the crystallization temperatures. These results indicate that the extent of liquid-liquid phase separation which exists below melting point is successfully controlled through the proper selection of solvent and thermal conditions.

Acknowledgements. We would like to thank Dr. M. Wolkowicz in Himont R&D Center for providing the polymer sample and L. G. Krauskopf in Exxon Chemical Co. for the solvents.

REFERENCES

1. Utracki, L. A. "Polymer Alloys and Blends", Hanser Publishers, **1989**.
2. Inaba, N.; Sato, K.; Suzuki, S.; Hashimoto, T. Macromolecules **1986**, *19*, 1690.
   Inaba, N.; Yamada, T.; Suzuki, S.; Hashimoto, T. Macromolecules **1988**, *21*, 407.
3. Schaaf, P.; Lotz, B.; Wittman, J. C. Polymer **1987**, *28*, 193.
4. Lee, H. K.; Myerson, A. S.; Levon, K. Macromolecules **1992**, *25*, 4002.
5. Cahn, J. W.; Hilliard, J. E. J. Chem. Phys. **1959**, *31*, 688.
6. Keith, H. D.; Padden, F. J., Jr. J. Appl. Phys. **1964**, *35*, 1270.
   Keith, H. D.; Padden, F. J., Jr. J. Appl. Phys. **1963**, *34*, 2409.
7. Cahn, J. W.; Hilliard, J. E. J. Appl. Phys. **1958**, *29*, 258.
   Cahn, J. W., J. Chem. Phys. **1965**, *42*, 93.
8. Van Aarsten, J. J. Eur. Polym. J. **1970**, *6*, 919.
9. Tanaka, H.; Nishi, T. Phys. Rev. Lett. **1985**, *55(10)*, 1102; Phys. Rev. A **1989**, *39(2)*, 783.
10. Lee, H. K. Ph.D. Dissertation, Polytechnic University, **1992**.

STRUCTURE AND THERMODYNAMIC ASPECTS OF PHASE SEGREGATION OF COMMERCIAL IMPACT POLYPROPYLENE COPOLYMERS

Francis M. Mirabella, Jr.

Quantum Chemical
Corporation
Process Research Center
Morris, IL 60450

INTRODUCTION

Commercial isotactic polypropylene homopolymer is widely employed in applications for thermoplastics. However, it is limited in its applications by its relatively low impact strength and high brittleness temperature. Its properties can be greatly enhanced by blending a minor fraction of a rubber–like material. This is commercially practiced by synthesizing reactor blends of isotactic polypropylene and ethylene–propylene rubbers (EPR) in–situ. Commercial blends of isotactic polypropylene and EPR are called impact polypropylene and constitute a significant segment of the polypropylene market. The effect of the added EPR is to toughen the brittle, low impact strength isotactic polypropylene matrix. The details of the toughening of a high Tg, hard polymeric material with a low Tg, rubberlike material are available elsewhere.[1] The rubber–like phase is known to exist as a phase–segregated discrete particle in a continuous matrix of the hard phase. The rubber–like particles are typically added at low concentration and, thus, form a discrete phase in the continuous matrix of the hard phase. An example of a toughened polymer is impact polypropylene[2]. Polypropylene itself has poor impact resistance and is very brittle at low temperatures. However, when about 10% – 20% of an ethylene/propylene copolymer (EPR) containing about 50% ethylene is added to polypropylene, the impact increases dramatically and the brittleness temperature is decreased markedly. The EPR is known to exist as very small, discrete particles in the polypropylene matrix[2]. These particles toughen the matrix against crack propagation by dissipating large amounts of energy in the matrix material around the particle, thereby blunting the crack and inhibiting crack propagation.[1]

Polypropylene–impact copolymers may be made by blending rubbers in polypropylene homopolymer or in–situ in a two–reactor system[3]. Typically commercial impact propylene is produced in–situ rather than by post–reactor blending and the superiority of this method is evidenced by the exceptional properties of the resins produced and their commercial success.[3] Polypropylene homopolymer is made in the first reactor and the EPR is made in a second reactor. The polypropylene is blended with the EPR in a sequential process in which the polypropylene is added into the second reactor while the EPR is being polymerized.

Impact polypropylene, produced in a sequential reactor system, is an extremely complex mixture of polypropylene homopolymers of varying stereoregularity and

New Advances in Polyolefins, Edited by
T.C. Chung, Plenum Press, New York, 1993

ethylene–propylene copolymers of varying composition. A complete molecular structure analysis of the impact polypropylene copolymer studied in this work will be given as a basis for understanding the primary focus of the study. This focus was the development of the final solid–state morphology of the impact polypropylene resin.

The final solid–state structure of commercial impact polypropylene copolymers is known to be multiphasic. However, there has been disagreement as to the mode of formation of the multiphasic system from the reactor powder. Specifically, there is uncertainty about the roles of liquid–liquid phase separation and crystallization of the matrix phase in the formation of the two–phase system. The major objective of this study was to determine the separate effects on the formation of the final multiphasic system of (a) demixing of the phases in the melt state and (b) crystallization of the semicrystalline, continuous polypropylene matrix phase.

EXPERIMENTAL

Analytical Temperature Rising Elution Fractionation (TREF)

Fractionation of the resins was done by temperature rising elution fractionation (TREF). This technique consists of dissolving the sample in trichlorobenzene (TCB) at a concentration of 0.007 g/mL at 140°C. This solution is deposited on a steel column (250 mm X 10 mm i.d.) packed with an inert support, Chromosorb P. The column is then capped and cooled to room temperature at 1.5°C/h over about 3 days. The column is then connected into a system through which TCB is pumped at 2.0 mL/min while the temperature is increased at 20°C/h. The species eluting from the column are detected with an IR detector (Miran 1A–CVF; Foxboro Analytical, East Bridgewater, MA) with a 1.5–mm pathlength, 4.5 µL internal volume, liquid flow–through cell with NaCl windows. It was set at a detection wavelength of 3.41 µm (C–H stretch). The eluting species can be trapped independently in fractions as a function of elution temperature. Only a very small mass in each fraction can be obtained in this primarily analytical technique. This technique is described in detail in ref. 4.

Preparative Temperature Rising Elution Fractionation

In order to collect large fractions of the fractionated resins, a preparative TREF apparatus was employed. The construction and operation of a similar version of this apparatus has previously been described in detail[5]. The polymer is dissolved in TCB and cooled slowly as in the analytical TREF procedure. The cooled polymer and solvent are then introduced into the chamber of the apparatus and heated slowly in incremental steps of temperature. Fractions are removed from a valve at the bottom of the apparatus. The polymer is recovered by evaporating the TCB solvent and drying thoroughly in a vacuum oven.

To facilitate the preparative scale fractionation, the soluble polymer, which was primarily the EPR, was removed as a room temperature–soluble prefraction in a preliminary fractionation step. The preliminary fractionation step was down as follows: The impact copolymer (PP–E–1) was dissolved at a concentration of 5.10 g in 250 mL of spectroscopic grade TCB. Solution was attained at approximately 160°C with gentle stirring. The solution was allowed to cool to room temperature and the precipitated polymer was recovered by centrifugation and filtration. The soluble polymer remained in TCB solution at room temperature. This procedure was repeated two additional times on the TCB insoluble polymer to exhaustively remove all soluble polymer. The soluble polymer was recovered by evaporative stripping of the TCB solvent and drying in a vacuum oven. The soluble polymer was 15.3 wt% of the original impact copolymer. Only the insoluble polymer fraction was introduced into the preparative fractionation apparatus. This technique is described in detail in ref. 4.

Thermal Analysis

The thermal behavior of the materials studied in this work was determined on a DuPont 990 differential scanning calorimeter (DSC). Samples of 5–10 mg were sealed in aluminum sample pans. The temperature was programmed at 10°C/min. Other parameters used were according to the typical DuPont 990 operating procedure.

^{13}C Nuclear Magic Resonance (NMR)

^{13}C–NMR was done on a JOEL FX–90Q NMR. Typical conditions were: 300 scans, 90°C pulse angle, 10–s pulse delay (ca. 1 h scan accumulation), and 10–mm tubes were used for relatively small sample mass (ca. 25–50 mg). Samples were swelled with a minimum volume (ca. 1 mL) of TCB. Data were handled according to standard techniques.

Polymer Processing

A one phase system, called a molecular dispersion, was obtained by dissolving the resin in xylene at 110–120°C and precipitating in ambient temperature methanol. The recovered materials were then dried and pressed at room temperature into opaque sheets.

Quiescent melt ripening of these materials was done at low pressure (effectively zero pressure on the press) in a hot press at a specified melt temperature for a specified length of time. After the specified time had elapsed, the samples were either quenched in ice water, or allowed to cool in the ambient air on the bench top. Specimens (15 mmx 15 mmx 0.9 mm) were sandwiched between aluminum foil (0.8 mm thick) sheets and aluminum back–up plates of 1/16" thickness. In the case of the shortest time specimen (5 seconds) the aluminum back–up plates were not used in order to facilitate rapid heat–up and cool–down.

The samples were taken from the hot–press and crystallized by quenching in about 2 seconds in ice–water or samples were allowed to cool on the bench–top and these required 5 minutes to crystallize. The time required for crystallization to occur in the bench–top cooled specimens was determined in two ways. A calibrated thermocouple probe was imbedded in a typical specimen. The specimen was put into the hot–press and rapidly reached 193°C. The specimen was removed from the hot–press and the time to reach 113°C was recorded. This was the crystallization temperature of polypropylene homopolymer, as determined by DSC. Also, the time–temperature data were plotted and the latent heat of crystallization was observed as a "flattened" (zero slope) portion of the curve. In both cases the time required for crystallization was 5 minutes.

The additional 5 min. in the melt state were added to the time in the melt of the bench–top cooled specimens.

The times chosen to hold the samples in the melt in the hot–press were 5 seconds, 45 seconds, 2 minutes, 8 minutes and 1 hour for the ice–water quenched specimens and 45 seconds, 2 minutes, 8 minutes and 1 hour for the bench–cooled specimens.

Electron Microscopy and Microtomy

The ripened specimens were microtomed at – 100°C in a Reichert FC4E ultra–microtome, etched in n–heptane at 60°C in a sonic bath for 20 minutes, ion coated with gold and analyzed in a JEOL 1200 EX electron microscope.

The electron photomicrographs were measured with an image analyzer. Particle areas were directly measured with the image analyzer. The areas were assumed to be circular and the corresponding calculated circle diameters were assumed to be the chord length of the intercepted particle. The required "true" particle diameter distribution moments ($\bar{d}n$, $\bar{d}w$ and $\bar{d}z$) were calculated from the chord lengths according to previously published statistical analysis methods of stereology[6]. The chord lengths obtained were converted to diameters of the three–dimensional structure with the following stereological formulae:

1.   Collect (li, Ni) data:

   li : chord length

   Ni : number of chords

2.   Calculate chord length averages:

   Harmonic Average: $\bar{l}_H = \dfrac{\Sigma\,N_i}{\Sigma\,N_i/l_i}$

   Number Average : $\bar{l}_n = \dfrac{\Sigma\,N_i\,l_i}{\Sigma\,N_i}$

   Weight Average : $\bar{l}_w = \dfrac{\Sigma\,N_i\,l_i^2}{\Sigma\,N_i l_i}$

3.   Calculate particle diameter averages:

   Number Average : $\bar{d}_n = \bar{l}_H \left(\dfrac{\pi}{2}\right)$

   Weight Average : $\bar{d}_w = \bar{l}_n \left(\dfrac{4}{\pi}\right)$

   Z–Average : $\bar{d}_z = \bar{l}_w \left(\dfrac{3\pi}{2}\right)$

   Calculate polydispersity, PD = $\bar{d}w/\bar{d}n$

Materials

The impact copolymer (PP/E–1) characterized in this work was a commercial impact polypropylene copolymer with a melt flow rate (MFR) of 2.5.

RESULTS AND DISCUSSION

A complete molecular structure analysis of a commercial impact polypropylene, produced in multi–reactor systems, has been rarely reported in the literature. The scarcity of this type of microstructural data on commercial impact–polypropylene resins is primarily due to the difficulty of separating the many different components of these complex mixtures. Identification and characterization of the components in the unfractionated whole polymer is presently not possible to accomplish, if a definitive analysis is required. The technique employed to overcome this difficulty was temperature rising elution fractionation (TREF)[4]. The analytical TREF elution pattern for the 2.5 MFR commercial impact polypropylene copolymer studied in this work is shown in Figure 1. This impact copolymer (PP/E–1) was fractionated in a preliminary step to remove the EPR component. The remainder of the polymer was fractionated by preparative TREF and separated into 12 fractions. The details of the characterization of the fractions has been given elsewhere.[7]

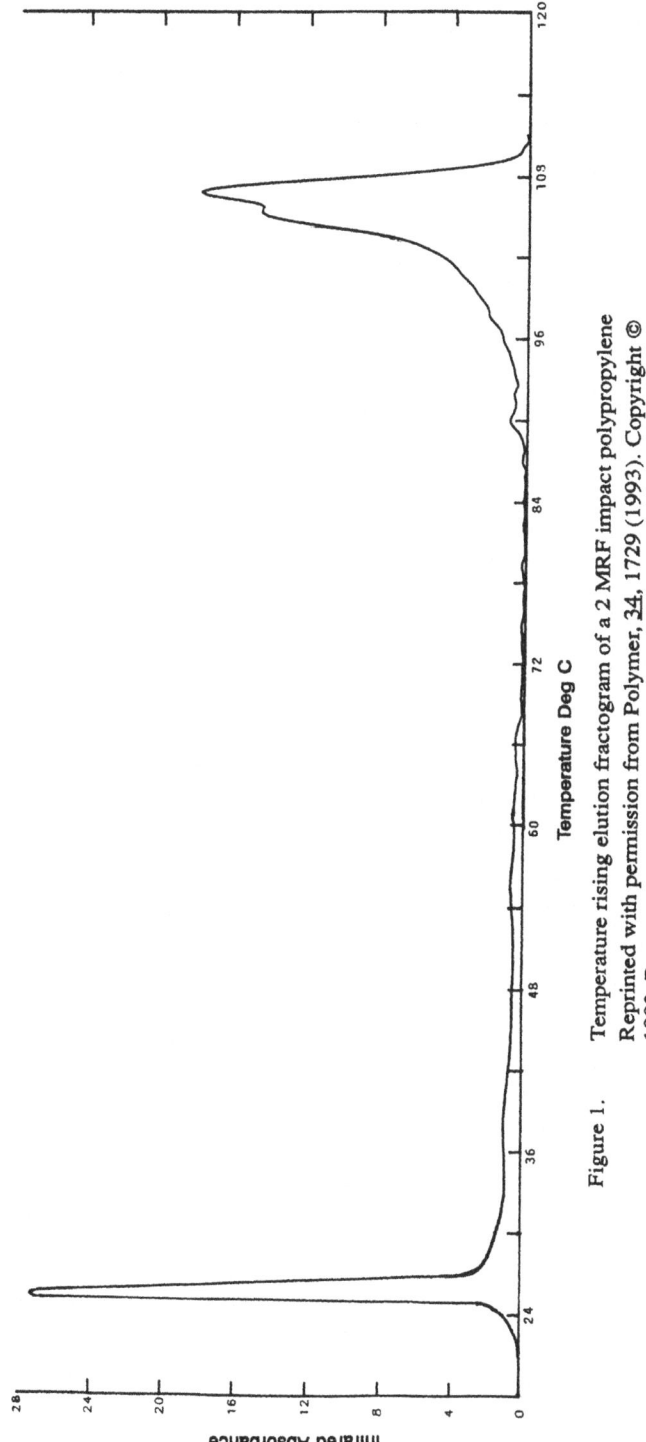

Figure 1.   Temperature rising elution fractogram of a 2 MRF impact polypropylene
Reprinted with permission from Polymer, 34, 1729 (1993). Copyright ©
1993, Butterworth–Heinemann Ltd.

TABLE 1

Components of Fractions from PP/E–1

Elution Temperature Range (°C)	Fraction Number	Fraction wt.%	Component Major	Minor	Trace
RT Prefraction		15.3	EPR 50/50 C₂/C₃	Atactic Polypropylene	
RT	1	0.4	E/P copolymers (negligible crystallinity)		
RT–40	2	1.4	E/P copolymers (negligible crystallinity)	PP[b]	
40–60	3	2.2	E/P copolymers (low crystallinity)	PP[b]	
60–85	4	3.5	LLDPE[c] + PP[b] (crystalline)	PP[b]	
89–90	5	2.1	LLDPE[c] + PP[b] (crystalline)	PP[b]	
90–95	6	1.4	PP[a] + LLDPE[c]	PP[b]	
95–100	7	1.6	PP[a]	PE (linear)[d]	PP[b]
100–105	8	6.1	PP[a]	PE (linear)[d]	
105–110	9	10.8	PP[a]	PE (linear)[d]	
110–115	10	23.1	PP[a]		PE (linear)[d]
115–120	11	13.2	PP[a]		PE (linear)[d]
120–140	12	18.9	PP[a]		PE (linear)[d]

[a] Highly Isotactic PP) The observation of high and low
) stereoregularity PP may be in
[b] Low Stereoregularity PP) different or the same molecular chain

[c] LLDPE refers to PE with a minor amount of propylene comonomer ($\leq 8\%$)

[d] PE (linear) refers to PE without observable comonomer.

The elution temperature ranges of the room temperature prefraction and the succeeding 12 preparative TREF fractions are presented in Table 1. These fractions were characterized by differential scanning calorimetry (DSC) and ^{13}C nuclear magnetic resonance spectroscopy (NMR)[7]. Table 1 shows the major and minor chemical components of these 13 fractions. In Table 1, PP = polypropylene, PE = polyethylene, and LLDPE = polyethylene with a small concentration of propylene. The complex structure of impact–propylene copolymers can be rationalized on the basis of the sequential polymerization process used to produce these polymers[7]. The weight percent of each fraction is given in Table 1. The polypropylene accounted for about 75% wt., the EPR about 17% wt. and the other ethylene–propylene copolymers about 8% wt. of the total.

The solid–state structure which has been observed for impact polypropylenes is biphasic and is basically composed of an isotactic polypropylene continuous matrix and an immiscible rubber–like EPR discrete particle phase[2]. The other components, i.e. the other E/P copolymers, are suspected to be incorporated into the EPR phase. This is based on miscibility considerations. There has been some disagreement about the genesis of the two phase system and the role of crystallization in the evolution of the segregated phases. This disagreement has arisen due to the overlapping in time of the processes of liquid–liquid demixing in the melt and the crystallization of the highly crystalline polypropylene matrix. The suggestion that the crystallization of the matrix has a controlling effect on the size and the shape of the dispersed phase has been made based on, for example, the early work of Keith and Padden. These workers showed that addition of low molecular weight "impurities" to high polymers had a controlling effect on spherulitic morphology[8], rates of crystallization [9] and interlamellar noncrystalline morphology[10]. These studies focused on the effects of the low molecular weight, noncrystallizing material on the crystallization process. However, these studies showed that the noncrystallizing species were segregated between lamellar crystals and between spherulitic boundaries. Therefore, by inverse reasoning it has been suggested that the noncrystalline material in multiphase systems is "squeezed out" by the advancing crystalline front and that its size, shape and position in the final dispersed phase system is controlled by the crystallization step. This section of the work was concerned with separating these two overlapping processes.

The synthesis of the impact polypropylene in–situ blend of components leaves open the difficult question as to the intimacy of mixing of the components in the reactor powder product. In order to avoid this difficult question for the purposes of this work the initial state of the polymer resin was artificially manipulated by dissolving in xylene at 110° – 120°C and precipitating in ambient temperature methanol. This process resulted in the production of an intimate mixture of the molecules which was called a "molecular dispersion." The molecular dispersion was one–phase with no macroscopic segregation of polypropylene homopolymer molecules and EPR molecules.

The resulting flocculated mass of polymer was dried thoroughly and cold–pressed into opaque sheets. The molecular dispersion specimens were then treated further to allow for the development of particle morphologies. This was accomplished by holding the specimens in a hot–press at low pressure in the melt at 193°C for specified time and then quenching in ice–water or cooling in the ambient air on the laboratory bench. Cooling from the melt on the bench until crystallization occurred took 5 min. for impact polypropylene. The time for cooling of the melt on the bench until crystallization occurred was added to the time in the melt in the press for these specimens. The quiescent processing in the melt state is called phase ripening or coarsening.[11]

The phase separation of immiscible molecules in polymer blends occurs rapidly in polymers melts due to a thermodynamic instability, this is called spinodal decomposition, or a thermodynamic metalstability, that is, nucleation and growth of a new phase.[12] However, this distinction of the character of the liquid–liquid phase transition is largely academic for the purposes of this work, because in either case the system rapidly produces nuclei of a new phase which are driven to "coarsen" into the final morphology by the same mechanism.[11]

The production of nuclei in these immiscible systems is, therefore, extremely rapid compared to the subsequent coarsening of the system to the final morphology. The short–time regime of coarsening is not understood and no theory exists for either the size or size distribution of the nuclei produced in the short–time regime[13]. The long–time regime of coarsening is reasonably well understood[13]. A simplistic rationale for the differing time scales of these two processes (nucleation and growth, and coarsening) can be developed as follows. At time zero the molecules of a binary solution, with molecule A being the major and molecule B being the minor component, are intimately mixed at the molecular level, i.e., there is no macroscopic phase segregation. This system has an extremely large driving force for segregation of the two types of molecules since the solution is highly supersaturated. Therefore, the molecules separate rapidly into two macroscopic phases. This rapid separation then effectively eliminates the supersaturation of the solution. In the next step of the process the minor phase, B, grows relatively slowly by combining of the nuclei of B. This occurs by the diffusion of molecules of B through the dilute solution of B molecules in the A matrix.

The driving force for this process is different from the first process. In this case, the B nuclei have extremely high surface area and, therefore, extremely high cumulative interfacial tension against the A matrix. the driving force is then to minimize the surface area, thereby minimizing the cumulative interfacial tension between A and B. This is done by simply increasing the particle size of B. The minimum of the interfacial tension is one particle of B in the A matrix.

The coarsening of a two–phase immiscible system has been kinetically modelled for the case of metals. The classical derivation was done following the theory of Ostwald ripening[11]. Again, the first step is very rapid, i.e. nucleation and growth, and is not included in the kinetic scheme of the model. The second step treats the slower growth of the grains of the minor phase in order to reduce the interfacial tension between the grains and the matrix. The results of this derivation for metals are most compactly expressed in the Lifshitz–Slyozov equation for a binary system[14]:

$$d^3 = d_o^3 + 64 \, \gamma \, (X_e \, V_m) \, Dt/9 \, RT \qquad (1)$$

where d_o = initial droplet diameter,
d = diameter at time t,
t = time,
γ = interfacial tension between the phases,
X_e = mole fraction of the minor phase in the matrix phase,
V_m = molar volume of the minor phase,
D = diffusion coefficient for the matrix phase
R = gas constant
T = absolute temperature

Based on this equation, it can be anticipated that certain variables will influence particle growth. The chemical composition of the copolymer phase will alter the interfacial tension and temperature will influence the diffusion coefficient.

The morphology of the melt–ripened specimens was determined by etching out the dispersed phase particles (EPR) and observing in a scanning electron microscope. The growth of the dispersed phase particles may be observed by consulting Figure 2 which shows SEM photomicrographs of the 5 second and 1 hour ice–water quenched specimens and the 45 second and 1 hour bench–cooled specimens. The diameter of the particles can be seen to be increasing and the number of particles per unit area can be seen to be decreasing with increased time in the melt. The particle diameters were estimated by measuring the chord length of a representative population of particles. It was found that particles were observed in the shortest time specimens (5 sec.). This was a confirmation of the rapidity of the nucleation and growth process. The particles grew larger with longer time in the melt. This is shown in Figure 3 in which the average particle diameter is plotted versus the cube root of the time according to equation 1. It can be observed in Figure 3 that the particle diameter increases linearly with $t^{1/3}$ for the ice water quenched specimens.

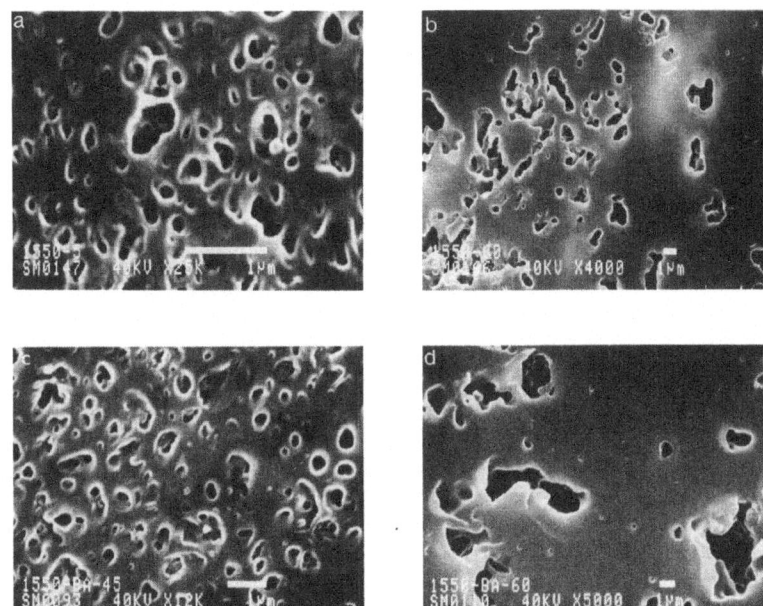

Figure 2.   Scanning Electron photomicrographs of n–heptane etched specimens.
Ice water quenched specimens:

    a.   5 seconds, 25,000 x magnification

    b.   1 hour,
            5,000 x magnification

Bench–cooled specimens:

    c.   45 seconds,
            1,200 x magnification

    d.   1 hour,
            4,000 x magnification

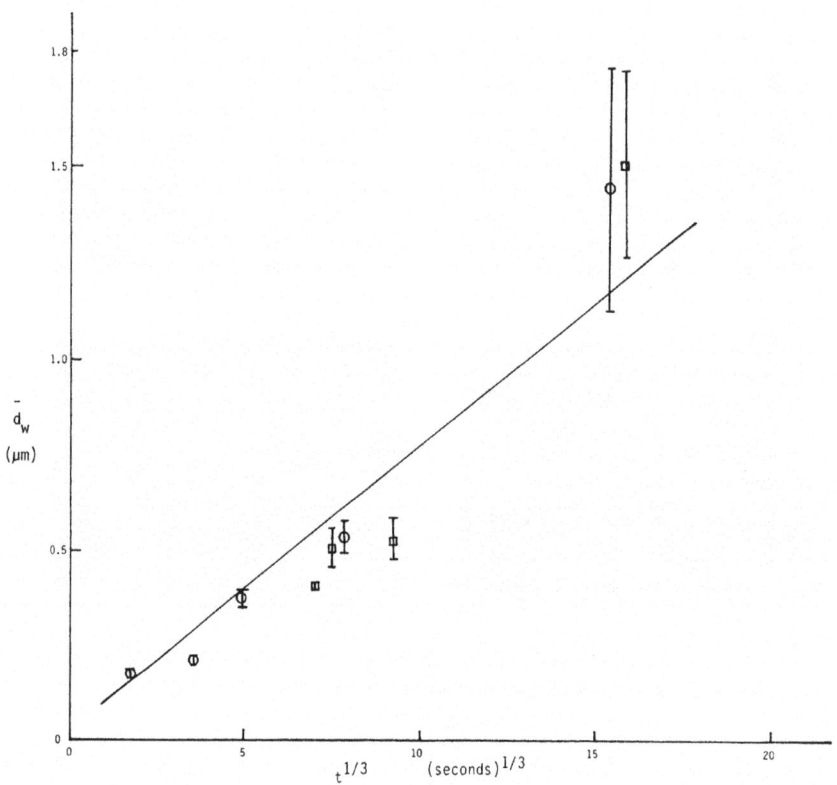

Figure 3.   Weight Average Diameter Versus the One–Third Power of the Time of Melt
Coarsening for Ice–Water Quenched and Bench–Cooled (Annealed)
Specimens.

Quenched from melt in ice water (193°C to 0°C)

Annealed on bench from melt (193°C to RT)

It can be further observed in Figure 3 that the bench–cooled specimens also fall along this line. The fact that the bench–cooled specimens fall along the line shows that it is the total time in the melt state that controls the growth of the phase segregated particles.

It may be observed in Figure 3 that the bench–cooled specimens that were held in the melt for shorter time (e.g. 45 seconds, 2 min. and 8 min.) fall below the line. This may be real and due to the cooling of the melt during the 5 min. of bench–cooling. As the melt cools the diffusion coefficient decreases, slowing the rate of coarsening. For the shorter time specimens this is expected because 5 min. is a significant period of time relative to the time these were held in the melt. However, this effect was not observed with the specimen held in the melt for 1 hour (see Figure 3). This may be expected because the 5 min. cooling on the bench is small compared to the 60 min. that this specimen was held in the melt.

A further observation was that no significant difference in the shape of the dispersed EPR particles existed between the ice–water quenched and bench–cooled specimens.

These results show that the variations in the cooling histories of the specimens had no effect on the phase segregation and the subsequent particle growth of the immiscible components. Further, the crystallization of the matrix had no observable effect on the rate of particle growth. The only effect of the crystallization process was to "lock–in" the particle morphology which had evolved in the melt state. The crystallizing front apparently grows past the dispersed phase particles as if these particles were not present, the particles having no effect on or participation in the crystallization.

CONCLUSIONS

The primary components of commercial impact polypropylenes are isotactic PP and EPR with minor components of atactic PP and other ethylene–propylene copolymers with higher crystallinity than EPR. The EPR component phase segregates due to liquid–liquid phase immiscibility and the phase morphology evolves in the melt state. The diameter of the EPR phase increases linearly with $t^{1/3}$, where t is the time in the melt state. The crystallization of the PP matrix phase has little or effect on the size or shape of the dispersed EPR phase, except to arrest or lock–in the coarsening of this phase at the time of crystallization.

REFERENCES

1. Bucknall, C.B., Toughened Plastics, International Ideas, Inc., Philadelphia, PA, 1977.

2. Prentice, P., Papapostolou, E. , and Williams, J. G., Polym. Mater. Eng. Sci., 51, 635 (1984).

3. Rifi, M. R., Ficker, H. K., and Walker, D. A., S.P.E. ANTEC, 316 (1986).

4. Wild, L. Advances in Polymer Science, 98, 1 (1990).

5. Kusy, R. P., and Whitley, J. Q., Biomed. J., Mater. Res., 20, 1373 (1986).

6. Weibel, E. R., Stereological Methods, Vol. 2: Theoretical Foundations, Academic Press, London/New York (1980).

7. Mirabella, F. M., Jr., Polymer, 34, 1729 (1993).

8. Keith, H. D., and Padden, F. J., Jr., J. Appl. Phys. 35 1270 (1964).

9. Keith, H. D., and Padden, F. J., Jr., J. Appl. Phys. 35 1286 (1964).

10. Keith, H. D., Padden, F. J., Jr. and Vadimsky, R. G., J. Polym. Sci., A–2, 4, 267 (1966).

11. Lifshitz, I. M., and Slyozov, V. V., J. Phys. Chem. Solids, $\underline{19}$, 35 (1961).

12. Binder, K., J. Chem. Phys., 79, 6387 (1983).

13. Voorhees, P. W., J. Stat. Phys., $\underline{38}$, 231 (1985).

14. McMaster, L.P., Adv. Chem. Ser., $\underline{142}$, 43 (1975).

INTERPHASE DESIGN IN CELLULOSE FIBER/POLYPROPYLENE COMPOSITES

P. Gatenholm and J.M. Felix

Department of Polymer Technology
Chalmers University of Technology
S-412 96 Göteborg, Sweden

Summary

True interphases were created in cellulose fiber/polypropylene(PP) composites by pre-treating the fibers with compatibilizing agents such as maleic anhydride modified polypropylene (MAPP) of different molecular weights. Titrimetric measurements confirmed the transesterification reaction between compatibilizers and cellulose and also allowed the estimation of compatibilizer segment lengths when these were fully stretched away from the cellulose surface. The lengths appeared to be proportional to the molecular weight of MAPP. Interphase thickness values in cellulose/PP composites, obtained by employing two different test methods, showed an increase in thickness with an increasing molecular weight of the compatibilizer. This indicates that the compatibilizer chains stretch away from the cellulose surface into the bulk, allowing them to interact with a larger fraction of the matrix molecules than when they are unstretched. The primary reasons for the stretching being thermodynamically favored are: Repulsion between the hydrophobic PP blocks of the MAPP and the hydrophilic cellulose substrate; the preference of the compatibilizer to be wetted by the PP matrix; and the relatively high grafting density of MAPP on the cellulose surface. The effect of stretching and interdiffusion on the adhesion between cellulose and PP was investigated using the peel test and the single-fiber fragmentation test. A directly proportional relationship was discovered between interphase thickness and adhesion. This, we believe, is a consequence of entanglements formed in an interdiffusion process at the fiber/matrix interface.

INTRODUCTION

The use of polyolefins as packaging materials has contributed to a rapid increase in the consumption of these plastics. At the same time, growing environmental awareness has turned much public attention to the plastic content in post-consumer waste, and has led to changes in legislation to promote recycling programs for polyolefins. There are unfortunately problems associated with the separation of the pure fraction of plastic in household waste, which limit material recycling. Such recycling would be of great value,

New Advances in Polyolefins, Edited by
T.C. Chung, Plenum Press, New York, 1993

however, as recent work has shown that the addition of wood cellulose greatly improves the performance of waste-borne polyolefins [1]. In view of this, a crucial need has been pointed out for research aimed to improve interactions in cellulosic-polyolefinic composites that would allow the development of new products from waste.

Poor compatibility between hydrophilic cellulose and hydrophobic polyolefins, leading to inferior properties of the waste-based composites, has called for some kind of adhesion-promoting step in the composite production. Polymers modified with maleic anhydride are commonly used as compatibilizing agents in polymeric multicomponent systems, and, for example, the graft polymer of polypropylene and maleic anhydride, MAPP, has demonstrated a remarkably positive effect on the mechanical properties of cellulose fiber/polypropylene composites [2-5]. Not only have the stiffness and tensile strength of the composites been improved, but the impact strength as well, which is unusual. The reason for this enhancement is believed to be related to changes in the interfacial region.

The intention of this study was to investigate the nature of the interphase and how it changes when compatibilizing agents of different molecular weights are used. Our interest is focused on the effect of the orientation of immobilized compatibilizer chains on the interphase thickness. The grafting density of the compatibilizer is determined by titrimetric measurements, and interphase thickness data are obtained according to two different procedures described elsewhere [6,7]. Adhesion between cellulose and PP was measured by peel testing and the single-fiber fragmentation test, and the effect of adhesion on mechanical properties of composites was evaluated by tensile testing.

EXPERIMENTAL

Materials

A cotton cellulose fiber from Parke-Davis was used in the single-fiber fragmentation test. Greaseproof paper, made of bleached sulphite pulp without additives, was used in the laminates evaluated in the peel tests. In all other experiments, α-cellulose fiber, originating from bleached sulphite pulp (Nymölla AB, Sweden), was used. After appropriate purification, i.e. Sohxlet extraction with suitable solvents, both fiber types had about the same surface composition, as obtained by Electron Spectroscopy for Chemical Analysis. Surface compositions after the treatments were also very similar for the different fiber types.

Three different compatibilizing agents were used for surface treatment of fibers: One alkenyl succinic anhydride and two polypropylenes of different molecular weights, grafted with the same monomer (maleïc anhydride). The alkenyl succinic anhydride, with a molecular weight of 350 g/mole, was commercially available (Ethyl Corp., USA), as were the two polypropylenes with MW of 39,000 (Hercoprime G, Hercules Inc., USA) and 4,500 (Epolene 43, Eastman Chemical, USA). Both polypropylenes were modified to contain a 6 weight-percent maleic anhydride.

The matrix material used in the composites was commercially available polypropylene with a molecular weight of 268,000 g/mole (Trespaphan NNA 30, Hoechst AG, Germany).

Methods

Cellulose treatments and titrimetric measurements were carried out as described elsewhere [8 and 3, respectively].

In the preparation of samples for the single-fiber fragmentation test, single cotton fibers, whose diameters and tensile strengths were measured by Scanning electron microscopy (Jeol JSM 5300) and tensile testing (Instron 1122), respectively, were placed between PP films which were melted together and then placed in a hot press for two minutes. Pressure was not applied during the first minute in the press, but a pressure of 30 kN was maintained during the second minute. Samples were then quench-cooled in ice water or allowed to crystallize completely at 130°C. Dogbone shaped specimens containing an aligned single fiber were punched out from the film.

Samples for peel tests were prepared as three-layer laminates in a hot press. A 60μm thick PP film was placed between two sheets of greaseproof paper, after which a pressure of 4 bar was applied for three min at 180°C.

The manufacturing of composites for tensile testing was carried out in two steps: The the filler, the additive and the matrix polymer were first mixed and homogenized in a mixing extruder (Buss-Kneader PR 46). The average residence time was in the order of 100 s and the temperature 180°C. The mixes were then injection-molded into tensile test bars (DIN 53455) at 180°C with a conventional injection molding machine (Arburg 221E/170R).

A miniature tensile tester (Minimat, Polymer Labs, Ltd.) was used for the single-fiber fragmentation tests. During these tests, the tensile tester was mounted under a polarizing light microscope (Olympus BH2-UMA) which allowed the observation of stress transfer and failure processes in the single-fiber samples. The fragmentation tests required a rate of 1 mm/min. Stress-strain data for composites was obtained by means of an Instron tensile tester (Instron 1193, extensometer G51-15MA). The strain rate was 1.1×10^{-3} s^{-1}.

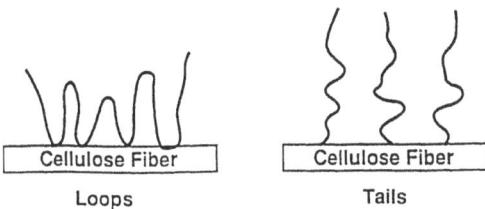

Loops   Tails

Figure 1. Schematic illustration of possible configurations of grafted cellulose.

RESULTS AND DISCUSSION

Surface configurations of grafted cellulose

In a previous paper, we showed that MAPP reacted with and became covalently bonded to the cellulose surface [3]. Titrimetric measurements were carried out in order to estimate the length of mobile chain segments of the compatibilizer (segments that are able to stretch away from the surface). Determinations were made of the extent of the transesterification, i.e. the number of carboxylic moieties on the compatibilizer that had reacted. Owing to the inhomogeneous composition of MAPP, we considered the two extreme cases: either all carboxylic groups that reacted were situated at the end of the PP chain or they were evenly distributed along the chain. These configurations are schematically shown in Figure 1.

In Figure 2, the theoretical lengths of the chain segments, when they are fully stretched away from the cellulose surface are shown as a function of the molecular weight of the compatibilizer. Possible lengths of fully stretched chains increase with increasing molecular weight. This fact would probably affect interphase thickness.

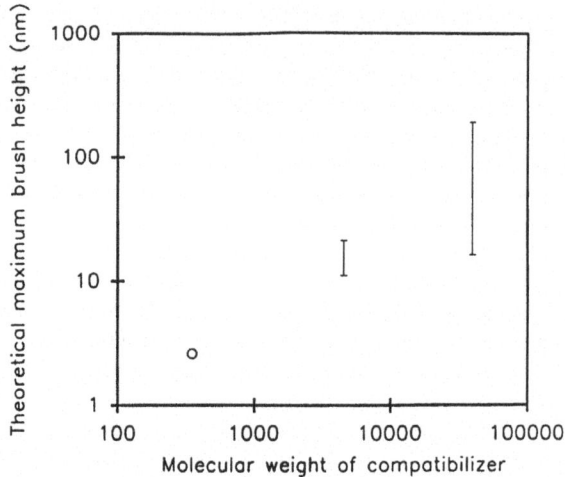

Figure 2. The effect of the molecular weight of polyolefinic compatibilizer on theoretical lengths of fully stretched chains.

Interphase structure

Several authors have indicated the interdiffusion between neighboring polymers resulting in entanglements to be crucial for bonding between phases [9,10]. Using dynamic mechanical measurements and tensile testing of composites, we have recently shown that the presence of compatibilizers considerably increased interphase thickness [6,7]. This effect was related to the molecular weight of the surface-attached compatibilizer: The thickest interphase was observed for the compatibilizer of the highest molecular weight.

If the lengths of the fully stretched compatibilizer chains, as determined by titrimetric measurements, are compared to the thickness of the interphases studied, we find that the interphases are always considerably thicker than the length of the compatibilizer chains. On the basis of these calculations, this means that the compatibilizer chains also restrict the mobility of matrix chains with which they are not in direct contact. The results indicate the probable stretching of compatibilizer chains away from the cellulose surface, thus yielding a brush-like interface [11] (see fig. 3).

This phenomenon should be thermodynamically favored by the following facts: A relatively high grafting density of compatibilizer on the cellulose surface causes the chains to stretch away to avoid overlapping; and the hydrophobic PP blocks of the compatibilizer prefer to

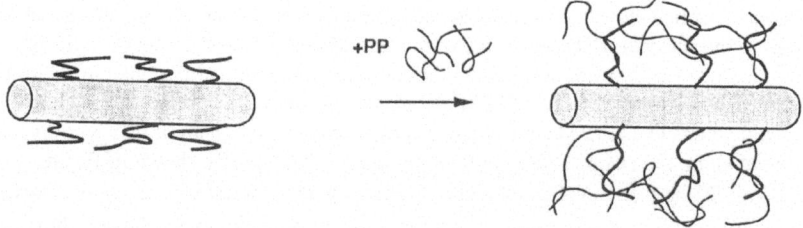

Figure 3. A model of the formation of a brush-like interface when a cellulose treated with MAPP is placed in a PP matrix.

be wetted by the PP matrix at the same time as they are repelled by the hydrophilic cellulose surface. The predominant factor in this case, we believe, is the effort of the hydrophobic compatibilizer chains to withdraw from the hydrophilic cellulose surface. In other words, the nature of the substrate would have a strong effect on the interphase in this system.

Adhesion in brush-like systems

In order to study the influence of the stretching of compatibilizer chains on adhesion between fiber and matrix, two methods were employed, namely peel testing and the single-fiber fragmentation test. The single-fiber fragmentation test is a very interesting method, as it yields not only interfacial shear strength data but also information on shear transfer and failure mechanisms at the fiber/matrix interface. During the fragmentation test, the load applied on the matrix is transferred to the fiber through shear at the interface. The stress in the fiber builds up until it breaks. This fragmentation process proceeds until no further fiber breakage occurs, i.e. until the critical fiber length is reached. The more efficient the interfacial shear stress transfer, the shorter the critical fiber length obtained. Detailed descriptions of the fragmentation test and the interpretation of results can be found elsewhere [12-14].

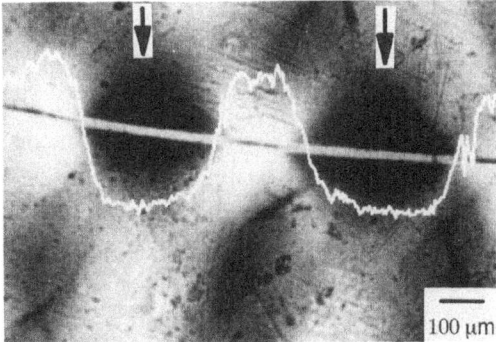

Figure 4. Birefringence patterns observed during tensile testing of a single cellulose fiber/LDPE composite. The white curve is the grey light intensity in the sample as obtained by image analysis of the micrograph. Arrows indicate areas where stress is efficiently transferred between fiber and matrix.

In the micrograph in Figure 4, it is shown how stress concentrations build up in a rayon fiber/polyethylene sample. In the areas in which the grey light intensity (the white curve obtained by image analysis) is high, the matrix carries the entire load, whereas, in the dark areas, the load is efficiently transferred to the fiber.

Two samples were tested using this method, both containing a fiber treated with the MAPP of the highest molecular weight. One of the samples was quench-cooled from the molten state, and the other was allowed to crystallize completely at 130°C. The interfacial shear strength data is shown in Table I.

Table I. Interfacial shear strength data as obtained for a cotton fiber/PP system using the single-fiber fragmentation test

Fiber type	Thermal conditions for the production of test specimen	Interfacial shear strength (MPa)
Cotton fiber treated with MAPP (MW=39,000)	Quench-cooled	18.6
	Isothermally crystallized (130°)	15.8

Apparently, the interfacial adhesion is stronger in the quenched sample than in the crystallized one. This may be a further indication of stretching in the melt. Quench-cooling may preserve the MAPP/PP entanglements formed in the melt, while slow cooling to some degree enables the exclusion of MAPP chains from the matrix phase when the spherulites are formed in the matrix. This mechanism is illustrated in Figure 5. Thus, the prevention of the existence of entanglements, formed by the stretching of MAPP into the melt, decreases interfacial shear strength.

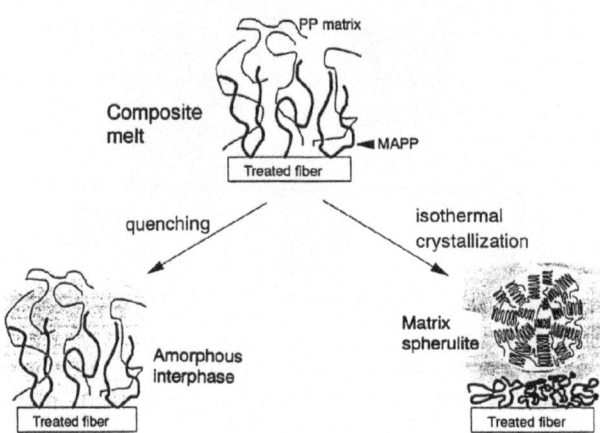

Figure 5. Interphase formation under different thermal conditions in a composite containing surface-grafted fibers.

Peel tests were performed only on quenched samples. This method has been used by other researchers to evaluate the effect of interdiffusion on adhesion [15]. In the present systems, we know that the surface properties (energetics, acid-base character etc.) of treated cellulose samples are rather similar [16]. Thus, when comparing the peel strength of these samples with PP, we consider any difference as primarily being a result of variations in the work required for disentanglement. Results of peel tests are presented in Figure 6. All treatments increased peel strength radically. For example, the use of the compatibilizer with the highest molecular weight increase peel strength by as much as three times. The interdiffusion of chains and the formation of entanglements at the interface are apparently phenomena of great significance in the interaction balance.

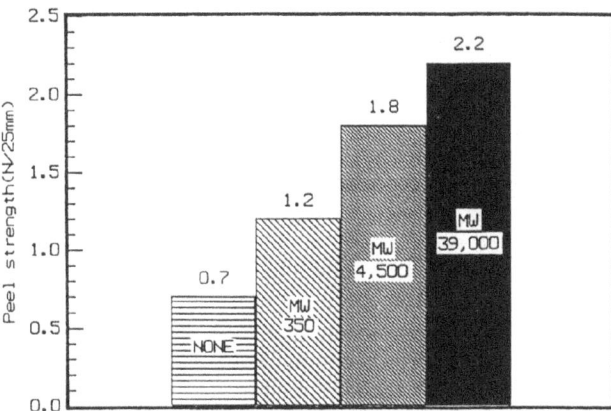

Figure 6. The effect of the molecular weight of the compatibilizer on the peel strength of cellulose/PP laminates.

The effect of the interphase on mechanical performance

The effect of the interphase structure on the mechanical performance of cellulose fiber/PP composites is illustrated in Figure 7, which shows the tensile strength of cellulose/PP composites as a function of fiber content and surface treatment.

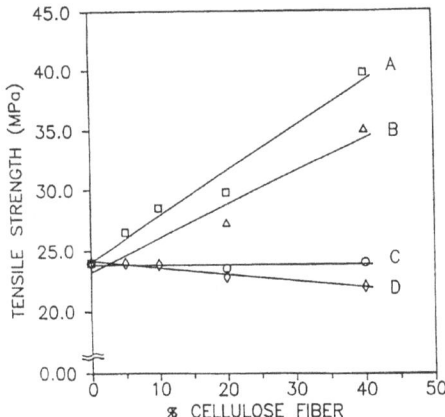

Figure 7. The effect of fibers and compatibilizer on the mechanical performance of composites. (D=sample with untreated fibers; A, B and C=samples with fibers treated with compatibilizer; molecular weights were 39,000; 4,500; and 350, respectively).

243

The introduction of untreated fibers into the PP matrix (sample D) results in a deterioration of strength, the effect being dependent on fiber content. Surface treatment of fibers with a low molecular weight compatibilizer (sample C) does not have a significant effect on the strength of composites. A very slight improvement in strength (as compared with the composite containing untreated fibers) can be seen for 40% fiber content. The addition of fibers still results in decreased composite strength. However, surface treatment with an MAPP compatibilizer of a 10 times higher molecular weight (sample B) yields a dramatic effect on tensile strength. This effect scales with fiber content. These fibers act as true reinforcing phases. The effect becomes even more pronounced when the molecular weight of the compatibilizer is increased further (sample A).

CONCLUDING REMARKS

The structure of the interphase has a crucial effect on the mechanical performance of cellulose/PP composites. A stretchable compatibilizer attached to the cellulose surface can diffuse into the PP matrix, providing entanglements which improve adhesion. Improved adhesion then makes the stress transfer more efficient, which in turn is of paramount importance for the reinforcement of PP by cellulose fibers.

ACKNOWLEDGEMENT

We gratefully acknowledge the financial support of the Swedish National Board for Technical Development.

REFERENCES

1. R.M. Rowell, T.L. Laufenberg and J.K. Rowell, "Materials Interactions Relevant to Recycling of Wood-Based Materials", Materials Research Society, Symposium Proceedings, Vol. 266, Materials Research Society, 1992.
2. R.G. Raj, B. V. Kokta, and C. Daneault, Intern. J. Polymeric M Mater., 12, 239 (1989).
3. J.M. Felix and P. Gatenholm, J. Appl. Polym. Sci., 42, 609 (1991).
4. G.E. Myers, I.S. Chahyadi, C.A. Coberly, and D.S. Ermer, Intern. J. Polymeric Mater., 15, 21 (1991).
5. S. Takase and N. Shiraishi, J. Appl. Polym. Sci., 37, 645 (1989).
6. J.M. Felix and P. Gatenholm, J. Appl. Polym. Sci., in press (1993).
7. J.M. Felix and P. Gatenholm, ACS Polym. Mater. Sci. Eng. Div. Proceedings, 67, Washington D.C., Aug 1992.
8. J.M. Felix and P. Gatenholm, in "Controlled Interphases in Composite Materials", H. Ishida, Ed., Elsevier, 1990, pp 267-276.
9. H.R. Brown, A.C.M. Yang, T.P. Russell, W. Volksen and E.J. Kramer, Polymer, 29, 1807 (1988).
10. S. Yukioka, K. Nagato, and T. Inoue, Polymer, 33, 1171 (1992).
11. S.T. Milner, Science, 251, 905 (1991).
12. R.K. Agrawal and L.T. Drzal, J. Adhesion, 29, 63 (1989).
13. W.D. Bascom, K.-J. Yon, R.M. Jensen and L. Cordner, J. Adhesion, 34, 79 (1991).
14. I. Verpoest, M. Desaeger and R. Keunings, in "Controlled Interphases in Composite Materials", H. Ishida, Ed., Elsevier, 1990, pp 653-666.
15. M.E. Fowler, J.W. Barlow and D.R. Paul, Polymer, 28, 2145 (1987).
16. J.M. Felix, P. Gatenholm and H.P. Schreiber, Polym. Eng. Sci., in press (1993).

INDEX

The manufacturer's authorised representative in the EU is Springer
Nature Customer Service Centre GmbH, Europaplatz 3, 69115 Heidelberg,
Germany. If you have any concerns regarding our products, please
contact ProductSafety@springernature.com

Printed and bound by CPI Group (UK) Ltd, Croydon, CR0 4YY
29/04/2026
02099472-0017